STRUCTURAL DYNAMICS FOR THE PRACTISING ENGINEER

TITLES OF RELATED INTEREST

Boundary element methods in solid mechanics
S. L. Crouch & A. M. Starfield

Computers in construction planning and control
M. J. Jackson

Earth structures engineering
R. J. Mitchell

Geology for civil engineers
A. C. McLean & C. D. Gribble

Geomorphological hazards in Los Angeles
R. U. Cooke

Hemispherical projection methods in rock mechanics
S. D. Priest

Numerical methods in engineering and science
G. de Vahl Davis

Plastic design
P. Zeman & H. M Irvine

Rock mechanics
B. H. G. Brady & E. T. Brown

Theory of vibration with applications
W. Thomson

STRUCTURAL DYNAMICS

FOR THE PRACTISING ENGINEER

Max Irvine

Department of Structural Engineering,
University of New South Wales

London
ALLEN & UNWIN
Boston   Sydney

Allen & Unwin (Publishers) Ltd,
40 Museum Street, London WC1A 1LU, UK

Allen & Unwin (Publishers) Ltd,
Park Lane, Hemel Hempstead, Herts HP2 4TE, UK

Allen & Unwin Inc.,
Fifty Cross Street, Winchester, Mass. 01890, USA

Allen & Unwin (Australia) Ltd,
8 Napier Street, North Sydney, NSW 2060, Australia

First published in 1986

British Library Cataloguing in Publication Data

Irvine, H. Max
  Structural dynamics for the practising engineer.
1. Structural dynamics
I. Title
624.1'71   TA654
ISBN 0-04-624007-1

Library of Congress Cataloging in Publication Data

Irvine, H. Max.
  Structural dynamics for the practising engineer.
1. Structural dynamics. I. Title.
TA654.178   1986   624.1'71   85-26805
ISBN 0-04-624007-1

Set in 10 on 12 point Times by
Mathematical Composition Setters Ltd, Salisbury, Wiltshire, UK
and printed in Great Britain by
Butler & Tanner Ltd, Frome and London

First of all one must observe that each pendulum has its own time of vibration, so definite and determinate that it is not possible to make it move with any other period than that which nature has given it. On the other hand one can confer motion upon even a heavy pendulum which is at rest simply by blowing against it. By repeating these blasts with a frequency which is the same as that of the pendulum one can impart considerable motion.

Galileo Galilei, *Discorsi a Due*
Nuove Scienze (1638)

Preface

This book has been written for practising engineers and senior under-graduates. It grew out of a set of notes prepared for classes given to each of these groups at different times of the academic year over the past several years. Both groups seem to enjoy the material and find that the applications are fairly direct.

Every engineering office or group needs people with a knowledge of the basic principles of structural dynamics but, regrettably, not every office has them. There is still a fairly widely held belief that structural dynamics is too difficult to be part and parcel of the structural engineer's technical arma-ment. It is true that it is difficult to find adequate time in the normal undergraduate programme to treat the subject in any detail: it may, therefore, be introduced as a technical elective at undergraduate level or find expression as a first-year graduate subject or as a continuing education course for the profession. However, like any other branch of engineering analysis, structural dynamics can be reduced to basic principles, and it is on those principles, properly applied, that many decisions can be made.

Thus, the aim of this book is to demonstrate that it is possible to get good information on the ramifications of the dynamic response of structural sys-tems without going into any great detail, and without employing advanced techniques of analysis. Nearly all systems can be reduced to equivalent single-degree systems and the quality of the resulting information is often entirely adequate for the purpose of decision making, given that a detailed knowledge of the system properties and loading is impossible anyway. Therefore, with the exception of the final chapter, where multi-degree-of-freedom systems are introduced, only single-degree-of-freedom systems have been considered.

The work is but an introduction. Accordingly, there is little that is new in the treatment, although some exceptions to this may be found in the numerical schemes described in Chapter 4. The book contains quite a lot of examples that have been worked in some detail. Many of these are drawn from practice. There are also exercises set at the end of each chapter. Where appropriate, reference has been made to some of the standard works in the field and a few papers and books dealing with specialised topics are cited, but the reference list is not intended to be exhaustive.

H. M. Irvine

vii

Acknowledgements

I gratefully acknowledge the assistance of Ruth Rogan and Colin Wingrove, who set the original script on the word processor. The figures were drawn by Mun Wye Yuen. Mario Attard, Ross Clarke, Glenn Dominish, Ray Lawther, Neil Mickleborough, Russell Staley, Weeks White and Peter Zeman read parts of the text and offered suggestions.

The style of the work owes much to the influence of the author's former colleague, John M. Biggs, Emeritus Professor of Civil Engineering at the Massachusetts Institute of Technology. To him and to Paul C. Jennings, Professor of Civil Engineering at the California Institute of Technology, I owe a debt of gratitude.

Figure 3.5 is reproduced by kind permission of Conoco (UK) Limited.

Contents

List of tables

1
Physical concepts

1.1 Conservation of energy and Newton's second law

Most loads applied to structures are dynamic in origin. That is, their manner of application and/or removal necessarily varies with time. Likewise, the response of a structure in resisting such loads has a temporal character.

Nevertheless, the distinction must be made between appreciable loads that are suddenly applied, or which oscillate about some particular loading level, and those whose full magnitude is reached only after a considerable delay. Examples of the latter (to give two extreme situations) are the filling of a reservoir behind a newly completed dam – the full lateral thrust will take months to develop – and the axial load levels in the ground-floor columns of a multistorey building as construction above ground proceeds. Obviously, these loads are essentially static.

But, as to the former, one has only to look at a flagpole to understand what is meant by movements that are generated by, say, sudden gusts of wind or by the periodic shedding of vortices from the pole itself. A gust that arrives after a period of quiescence causes the pole to flex suddenly and vibration ensues, in which the displacement varies periodically, the velocity changes sign periodically and, by and large, peak values of the one correspond to zero or mean values of the other. Thus, in adjusting to changed circumstances, the pole acquires kinetic energy (on account of its velocity) and stores strain energy (on account of its displacement).

In the absence of dissipative agents (such as damping), the statement of work-energy conservation is

$$KE + PE = W \tag{1.1}$$

where KE is the kinetic energy acquired by the vibrating system, PE is the potential or strain energy stored and W is the work done by the external forces.

This equality between internal energy storage and work done must be maintained at every point in time after the application of the external forces. Hence

$$\frac{d}{dt}(KE + PE - W) = 0 \tag{1.2}$$

1

The time derivative of the energy balance is also zero, for if it was not Equation 1.1 could not, in general, be true. However, a cautionary note must be sounded regarding evaluation of the quantity dW/dt when the external forcing is time varying. But, since this does not create a problem with any example in this chapter, the matter is relegated to Section 1.4.

Newton's second law states that the time rate of change of momentum of a body is equal to the impressed force. In all cases of interest in this book the mass of the system is constant and the rate of change of momentum is simply the product of mass and acceleration. Supposing the external force to be known, the impressed force is equal to this force minus the force arising from the internal resistance offered to movement. This last force component may be derived from structural theory and, again, in the absence of frictional forces, such as damping, is clearly related to the product of stiffness and displacement.

Thus, force equilibrium requires that

$$M\frac{d^2\Delta}{dt^2} = P - K\Delta$$

or

$$M\frac{d^2\Delta}{dt^2} + K\Delta = P \tag{1.3}$$

where P is the external applied force, K is the structural stiffness of the system and M is its mass. The acceleration and displacement terms are $d^2\Delta/dt^2$ and Δ, respectively.

Equation 1.3 is a second-order ordinary differential equation, which can in theory be solved for any value of P provided that two initial conditions, or starting values, are specified. It is usual for initial values of the response displacement Δ and response velocity $d\Delta/dt$ to be given. However, to talk about a solution at this stage is somewhat premature.

The trouble with the above differential equation of motion is that it is really not very useful in its present form. Consider the case of the flagpole alluded to earlier. To be sure, there is little to quibble over in taking the pole-tip deflection as representative of Δ. But what value of stiffness K do we associate with it? Furthermore, we cannot use the total mass of the pole when writing the inertia term $M\,d^2\Delta/dt^2$ since, clearly, not all the pole has the same acceleration as its tip. Even more basic is the fact that, while we may well have a fairly detailed knowledge of the way in which gust loading is distributed with height, its sum is not the appropriate term to use as the external force acting in this case.

A vibrating pole is one of the more straightforward problems in structural dynamics, and yet we see that we are in a quandary as to how to describe its dynamic equilibrium quantitatively. This is where a knowledge of energy principles is invaluable. In the case of the flagpole, if we are able to establish, approximately, the profile of deflection with height (and in this respect any smooth curve that pays lip service to the nature of the loading and the boundary conditions would do), we can in turn find expressions for the relevant entries in the statement of conservation of energy. This would then allow P, K and M to be established.

However, we are not adding extra information by considering both force equilibrium and energy conservation: one statement is a consequence of the other. The point is that an energy approach frequently provides a convenient framework in which problems of dynamic equilibrium may be set up. In the practice of structural dynamics, setting up the equation of motion is a vital adjunct to its subsequent solution.

Consequently, the thrust of the present chapter is to illustrate, by way of an example or two drawn from practice, how one may go about setting up (and solving) the sorts of problems that could come across the practising engineer's desk. The purpose is to sketch in a line of approach that is firmly rooted in fundamentals. Necessarily, approximations consistent with engineering judgement are required, but the line of approach is widely applicable.

So in Section 1.2 we discuss free vibrations, and in Section 1.3 we spend some time on response to a suddenly applied load. In each case, conservation of energy is the starting point.

1.2 Free vibrations

The spring–mass system shown in Figure 1.1 is an idealisation of some structure. Notwithstanding our earlier reservations, we suppose that the correct mass M has been found, as has the required stiffness K, and that the displacement Δ is representative of some point in the structure.

We are therefore making the assumption that the structure can be modelled as a single-degree-of-freedom oscillator, and that degree of

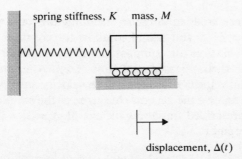

spring stiffness, K mass, M

displacement, $\Delta(t)$

Figure 1.1 Definition diagram for a single-degree-of-freedom oscillator.

freedom is the displacement Δ. A considerable amount of useful information can be gained from such simplified physical models of structural systems, provided that the structure does not have glaring discontinuities or ambiguities in the distribution of masses and local stiffnesses. In such cases it is more difficult to decide, *a priori*, on a single representative displacement, and additional displacements – that is, more degrees of freedom – may be necessary. For instance, a dynamic analysis of an elevated water tank in which sloshing is deemed important would probably require at least two degrees of freedom – lateral movement of the tank and vertical sloshing motion of the water. In this book we are for the most part restricting ourselves to single-degree systems because this is, by and large, the most fertile ground for the practitioner.

Returning to our spring–mass system, we note that in the absence of an external force and in the absence of damping, Equation 1.2 becomes

$$\frac{d}{dt}(KE + PE) = 0$$

or

$$KE + PE = \text{constant} \tag{1.4}$$

The kinetic energy is

$$KE = \tfrac{1}{2}M\left(\frac{d\Delta}{dt}\right)^2 = \tfrac{1}{2}M\dot{\Delta}^2$$

and the potential energy is

$$PE = \tfrac{1}{2}K\Delta^2$$

so Equation 1.4 becomes

$$\tfrac{1}{2}M\dot{\Delta}^2 + \tfrac{1}{2}K\Delta^2 = \text{constant} \tag{1.5}$$

This energy balance holds for all time, including the start, where we are free to stipulate values of Δ and $\dot{\Delta}$ and so allow the constant to be evaluated.

To start the vibration off, suppose the mass is pulled outwards by an amount Δ_0 and then released (see Fig. 1.2). Alternatively, we could prescribe an initial velocity or, indeed, both velocity and displacement could be prescribed initially – the general character of the solution is unchanged.

Hence, with prescribed initial conditions of $\Delta = \Delta_0$ and $\dot{\Delta} = 0$ at $t = 0$, Equation 1.5 becomes

$$\tfrac{1}{2}M\dot{\Delta}^2 + \tfrac{1}{2}K\Delta^2 = \tfrac{1}{2}K\Delta_0^2$$

4

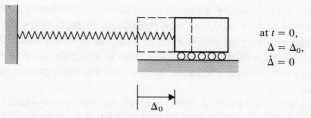

Figure 1.2 Initial conditions – mass pulled sideways by Δ_0 and released.

This equation may be rendered dimensionless by dividing both sides by $\frac{1}{2}K\Delta_0^2$, which is the total energy of the system in this instance. The result is

$$\left(\frac{\dot{\Delta}}{\sqrt{(K/M)}\,\Delta_0}\right)^2 + \left(\frac{\Delta}{\Delta_0}\right)^2 = 1 \tag{1.6}$$

This is the equation of a circle of radius unity (i.e. it is of the form $y^2 + x^2 = 1$).

Plotting Equation 1.6 as in Figure 1.3a, and labelling the path A, B, C, D, allows the following points to be made.

Position A corresponds to conditions at $t = 0$ when $\Delta = \Delta_0$ and $\dot{\Delta} = 0$. Subsequently, the potential energy reduces, the reduction being compensated for by a gain in kinetic energy because the mass starts to move back towards the position in which the spring is unstretched. The velocity is negative and the path followed is from A to B. At B all the potential energy has been converted to kinetic energy: the spring is unstretched but the mass is moving with its greatest velocity, which is of magnitude $\dot{\Delta}_{max} = \sqrt{(K/M)}\Delta_0$.

Continuing on past B, the mass starts to slow down as the spring is compressed. The kinetic energy acquired is now being transferred to potential energy, and when point C is reached the mass is at the mirror image position relative to A. The path from C to D mirrors that from A to B and the loop is closed over the last quarter, D to A.

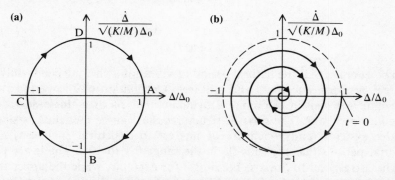

Figure 1.3 Velocity–displacement diagram for free vibrations: (a) without damping; (b) with damping.

The process then starts over again and will, in theory, repeat itself indefinitely because we have ignored damping. In fact, some damping will always be present, so that at the start of each subsequent cycle there is less energy available to continue the motion. Not unreasonably, this progressive decay in free vibrational response is depicted by the concentric spiral of Figure 1.3b.

The representative unit of time insofar as the spring–mass system is concerned is the time to traverse the loop ABCDA. Because of the radial symmetry of the velocity–displacement curve, the only functions of time that maintain the repeatability we require are the circular trigonometric functions (in view of Figure 1.3 it is obvious why the adjective 'circular' is used). Thus, the most general solution to Equation 1.6 that we could imagine is

$$\Delta(t) = A \sin\left(\frac{2\pi t}{T}\right) + B \cos\left(\frac{2\pi t}{T}\right)$$

where A and B are constants and T is the representative unit of time. The chosen initial conditions of $\Delta(0) = \Delta_0$, $\dot{\Delta}(0) = 0$ mean that $A = 0$ and $B = \Delta_0$. Hence the specific solution is

$$\Delta(t) = \Delta_0 \cos\left(\frac{2\pi t}{T}\right) \tag{1.7}$$

Substitution of this in Equation 1.6 yields

$$\left(\frac{2\pi}{T\sqrt{(K/M)}}\right)^2 \sin^2\left(\frac{2\pi t}{T}\right) + \cos^2\left(\frac{2\pi t}{T}\right) = 1$$

and the familiar trigonometric identity, $\sin^2\theta + \cos^2\theta = 1$, can be preserved only if

$$\frac{2\pi}{T\sqrt{(K/M)}} = 1$$

or

$$T = 2\pi\sqrt{(M/K)} \tag{1.8}$$

The quantity T is the natural period of vibration, which depends only on system properties of mass and stiffness. Calculation of the period is an essential first step in beginning a dynamic analysis of a single-degree-of-freedom oscillator. The sorts of structures that can be modelled as single-degree systems (and which are of interest to structural engineers) have natural periods that typically lie in the range 0.1 to 10 s. The lower limit might correspond to low-rise buildings, for example, while the upper limit could correspond to very tall buildings or long-span bridges – the Golden Gate Bridge (a suspension bridge) has a fundamental period of about 8.5 s.

In recent years there has been some interest in structures that are exceptional as regards length of their fundamental periods. In Chapter 3 we consider a tension-leg platform – an offshore structure that can be classed as a giant inverted pendulum – which has a period of about 1 min.

In structural systems potential energy can be stored either as gravitational potential energy (as in the pendulum) or as elastic strain energy (that is, Hooke's law is invoked). It is generally the case that, when gravity is the primary restoring force, the structure can be classed as compliant or flexible.

In itself, flexibility is not necessarily an undesirable characteristic. There are situations, as we will see later, when the ability to 'give' is highly desirable. However, when it impinges directly on user comfort – cases in point being slender high-rise buildings, footbridges and walkways on road bridges – there can be problems. Static serviceability criteria that govern live-load deflections do so by, in effect, requiring adequate structural stiffness. On the other hand, a dynamic criterion of adequate stiffness under live loads is that the natural period be reasonably low. This requires both high structural stiffness and low mass associated with the dead weight of the structure. Unless care is taken, these dual requirements can be mutually exclusive.

Returning to Equation 1.8 we note that the mass M has units of force $\times$ time^2/length, while the stiffness K has units of force/length. Consequently, the equation is dimensionally correct. A common mistake in calculating periods is to forget to convert from weight to mass (by dividing by the acceleration due to gravity).

The unit of time for the natural period is seconds. The inverse of the period is the frequency f, where

$$f = \frac{1}{T} = \frac{1}{2\pi} \sqrt{(K/M)} \tag{1.9}$$

Its units are cycles per second or, the equivalent, hertz (Hz). Another definition of frequency is also common and that is the circular frequency ω, which is

$$\omega = 2\pi f = \sqrt{(K/M)} \tag{1.10}$$

Since there are 2π radians in a circle, the units of ω are radians per second. The natural circular frequency is often a convenient quantity to use when dynamic analyses are being undertaken.

Finally, note that any point on the circle depicted in Figure 1.3a has co-ordinates of vibrational displacement and velocity, respectively, and that the angle swept out by passage around the curve is ωt. Consequently, the co-ordinates of the point are $\Delta_0 \cos \omega t$, $-\omega \Delta_0 \sin \omega t$.

7

Example 1.1   Impact loads on an offshore crane

A rather crude representation of an offshore crane is shown in Figure 1.4. The purpose of the example is to establish simple design criteria for the two basic types of crane movement – slewing and hoisting.

SLEWING

Slewing involves rotation of the crane as a whole about a vertical axis more or less through A. That is, the payload is being transported horizontally. Sudden slewing motions will generate pendulum motion of the hoist BC, causing a horizontal force perpendicular to the plane of the page to be transmitted to the tip of the boom AB. This invites the possibility of a substantial bending moment at the base of the boom.

Figure 1.5a shows an end sectional view of the boom tip and the hoist rope. Consider free vibrations of the pendulum consisting of payload and hoist rope, it being assumed that the boom is rigid. This assumption is reasonable given the flexibility inherent in pendulum action.

The kinetic energy of the payload is

$$\text{KE} = \tfrac{1}{2}M(L_1\dot{\theta})^2 = \tfrac{1}{2}M\dot{\Delta}^2$$

since $(L_1\dot{\theta})$ is the velocity of the payload because θ is the angle the hoist subtends to the vertical. The gravitational potential energy acquired is the difference between two quantities, namely MgL_1, the reference value, and $MgL_1 \cos\theta$, the value in the displaced position. That is

$$\text{PE} = MgL_1(1 - \cos\theta)$$

Supposing θ to be small, we may use the small-angle approximation of $\cos\theta \simeq 1 - (\theta^2/2)$ and write

$$\text{PE} \simeq \tfrac{1}{2}MgL_1\theta^2 = \tfrac{1}{2}(Mg/L_1)(L_1\theta)^2 = \tfrac{1}{2}K\Delta^2$$

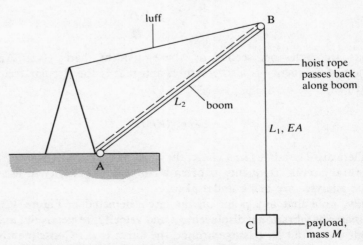

Figure 1.4 Schematic of an idealised offshore crane.

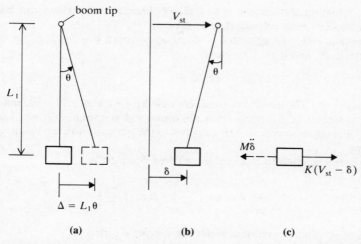

Figure 1.5 Pendulum action in the hoist rope: (a) free vibrations; (b) slewing; (c) horizontal equilibrium.

and the stiffness in this case is purely gravitational, namely $K = Mg/L_1$.

The period of the motion is thus

$$T = 2\pi \sqrt{\left(\frac{M}{K}\right)} = 2\pi \sqrt{\left(\frac{L_1}{g}\right)}$$

and we establish the curious fact that the period of a simple pendulum is independent of the mass on account of the appearance of a mass term in the stiffness. Galileo knew this from his experiments and it was Huygens who supplied the necessary theory somewhat later. Huygens' period formula was probably the first devised by mathematics.

The situation in Figure 1.5b is that the boom is rotated about its base with a constant angular velocity that gives rise to a velocity V_s, the slewing speed of the boom tip. Initially the payload will tend to lag the boom tip and this inclination to the vertical gives rise to the lateral shear that the boom will experience as a result of a sudden slewing motion. After a time t the boom tip has moved a distance $V_s t$ while the absolute movement of the payload is δ where

$$\delta = V_s t - L_1\theta = V_s t - \Delta$$

From the free-body diagram of Figure 1.5c, the inertia force is based on the absolute acceleration of the mass, namely, $M\ddot{\delta}$, while the spring force is based on the relative displacement $(V_2 t - \delta)$, namely $K(V_s t - \delta)$. In showing the inertia force dotted in the free-body diagram we have invoked D'Alembert's principle by which dynamic equilibrium can be treated in an analogous way to static equilibrium. We will do this repeatedly throughout the text. Newton's second laws states

$$M\ddot{\delta} = K(V_s t - \delta) \qquad \text{or} \qquad M\ddot{\delta} + K\delta = K V_s t$$

9

and the initial conditions are $\delta(0) = \dot{\delta}(0) = 0$ because the payload can have no absolute movement or velocity right at the start.

Rather than solve this equation, it is simpler to couch it in terms of the relative displacement Δ, and it then becomes

$$M\ddot{\Delta} + K\Delta = 0$$

because $\ddot{\delta} = -\ddot{\Delta}$. The initial conditions are now $\Delta(0) = 0$ and $\dot{\Delta}(0) = V_s$, and so the prescribed motion of the boom tip is physically and mathematically equivalent to an equal and opposite velocity being applied to the payload, with the boom tip held stationary.

The general solution to this problem of free vibration is

$$\Delta(t) = A \sin\left(\frac{2\pi t}{T}\right) + B \cos\left(\frac{2\pi t}{T}\right)$$

and use of the initial conditions yields the specific solution

$$\Delta(t) = \frac{V_s T}{2\pi} \sin\left(\frac{2\pi t}{T}\right) = V_s \sqrt{\left(\frac{L_1}{g}\right)} \sin\left(\frac{2\pi t}{T}\right)$$

The peak lateral load occurs at a time $T/4$ after the start and is

$$Mg(\Delta_{max}/L_1) \qquad \text{or} \qquad MgV_s/\sqrt{(gL_1)}$$

HOISTING

Hoisting involves the vertical transportation of the payload by winching the hoist rope. This, then, is the other fundamental type of crane movement.

For simplicity we shall assume that all members in Figure 1.4, with the exception of the hoist rope that passes along the boom and over a pulley at B, are rigid. This assumption can easily be relaxed, but only at the expense of simplicity. Our physical model is then as shown in Figure 1.6.

The stiffness is

$$K = EA/(L_1 + L_2)$$

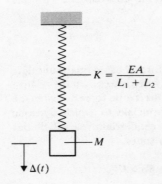

$$K = \frac{EA}{L_1 + L_2}$$

M

$\Delta(t)$

Figure 1.6 Single-degree-of-freedom model of hoisting operation.

10

where EA is the product of Young's modulus for the steel rope and its equivalent cross-sectional area. The operative mass is simply that of the payload, M.

There is a potentially severe event that must be designed against, as follows. In an offshore platform a tending vessel brings the payload alongside and the crane is used to hoist it aboard. The vessel can drop quite rapidly on a wave and in those circumstances the loading is derived from the tending vessel dropping downwards with the winch simultaneously winding in the hoist rope.

In the context of a vibrational study we are, in effect, gradually applying a load whose mass is that of the spring–mass system. The initial velocity V_h, which is the algebraic sum of the two effects mentioned above, is constant until the spring has stretched sufficiently to carry the payload Mg. Thus, initially, the payload must move with the vessel. After this, the payload lifts clear of the tending vessel and further stretching occurs because of the kinetic energy acquired by the payload.

The hoist rope continues to stretch until the velocity of the mass is momentarily zero. This corresponds to the greatest possible spring extension and, therefore, the greatest axial force in the hoist rope. If we call this peak extension Δ_{max}, the additional strain energy stored after the payload first lifts clear of the tending vessel is

$$\tfrac{1}{2}K\Delta_{max}^2 - \tfrac{1}{2}K(Mg/K)^2$$

where Mg/K is the stretch in the hoist rope when the weight Mg is just taken up.

The energy that has to be transferred to give this extension comes from two sources: there is the kinetic energy imparted, and there is the loss in gravitational potential energy. Thus

$$\tfrac{1}{2}MV_h^2 + Mg[\Delta_{max} - (Mg/K)] = \tfrac{1}{2}K\Delta_{max}^2 - \tfrac{1}{2}K(Mg/K)^2$$

This is a quadratic in Δ_{max} and the required root is

$$\Delta_{max} = \frac{Mg}{K}\left[1 + \sqrt{\left(\frac{KV_h^2}{Mg^2}\right)}\right]$$

Therefore, the peak dynamic axial force in the rope is

$$K\Delta_{max} = Mg[1 + (\omega V_h/g)]$$

where $\omega = \sqrt{(K/M)} = \sqrt{\{EA/[(L_1 + L_2)M]\}}$, the natural frequency for axial motion of the hoist rope–payload system. Relating this to the static load to be carried gives rise to a dynamic amplification factor (DAF)

$$\mathrm{DAF} = \frac{K\Delta_{max}}{Mg} = 1 + \frac{\omega V_h}{g}$$

This is the Det Norske Veritas formula for offshore cranes. Problem 2.7 at the end of Chapter 2 indicates another route to the solution.

The DAF is always greater than unity – the payload must be supported – and can be substantially greater if the velocity on the wave is high, for example. Engineers working with structures that may be subjected to dynamic loads spend a good deal of time attempting to estimate DAFs, for reasons that are fairly obvious.

In terms of numbers, a large offshore crane might, at minimum boom radius, be required to lift 500 kN. A typical value of EA (allowing for falls, etc.) would be about 5×10^5 kN, while $L_1 + L_2$ would be of the order of 100 m. Suppose the prescribed initial velocity is $V_h = 2$ m s^{-1} – the majority of this being due to heave of the tending vessel. The acceleration due to gravity is 9.81 m s^{-2}. The dynamic amplification factor is

$$\text{DAF} = 1 + \sqrt{\left(\frac{5 \times 10^3 \times 9.81}{500}\right) \times \frac{2}{9.81}} = 3.02$$

In this example the largest DAF, when combined with its payload, may not give rise to the largest dynamic effects in the crane because of the interplay between boom radius (in plan) and associated 'safe' payload. Several cases must be investigated. A more detailed examination is made in Example 5.7.

Example 1.2 *Design criterion for the chain of a single-point mooring*

This example also involves the interplay of kinetic energy and potential energy, but in a different way from that considered in the previous example.

A single-point mooring sometimes consists of a heavy chain anchored to a 'dead man', or pile cap, at the ocean floor and attached at its upper end to a buoy floating on the surface. The vessel berths at the buoy (see Fig. 1.7).

Under static design conditions, which consider wind drag and current drag on the vessel due to a running tide, it is a straightforward matter to estimate the size of chain necessary safely to resist the drag force. However, the matter cannot be laid to rest for the following reason.

Changes in tidal current and wind can occur abruptly. If both were to happen, say reversing completely, the vessel would start to drift towards the centre of its mooring circle. By the time it passes through the centre it will have acquired a small velocity – of the order of 0.5 m s^{-1} say – and the chain will hang limp and vertical. Continuing on, the chain will begin to pick up and tighten. But, by the time the

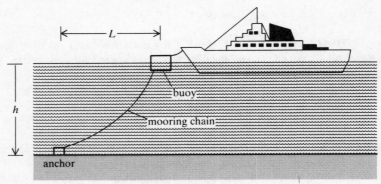

Figure 1.7 A single-point mooring.

vessel is close to the other side of the mooring circle, there is no guarantee that it will have stopped. In fact, when the chain does finally tighten, the jerk transmitted could very well break the chain because, even though the velocity of the vessel was small, its kinetic energy was huge.

The actual picture, instant by instant, is quite complicated, but the great advantage of an energy approach is that we can consider conditions before and after and let matters in between times take care of themselves.

The energy picture is as follows. At the centre of the mooring circle the vessel has acquired kinetic energy of

$$\text{KE} = \tfrac{1}{2}MV^2$$

where M is the mass of the vessel and V its velocity. The gravitational potential energy of the chain is then

$$\text{PE} = (mgh)h/2$$

where mg is the submerged weight per unit length of the chain and h is the water depth.

The actual chain length is approximately $\sqrt{(L^2 + h^2)}$, where L is the radius of the mooring circle. In general, a good portion of the chain lies on the sea bed when the vessel is at the centre of the mooring circle.

In order for the vessel to come to rest at the other side of the mooring circle, the total energy (KE + PE, as above) must be stored entirely as gravitational potential energy in the fully extended chain. This potential energy is

$$\text{PE} = [mg\sqrt{(L^2 + h^2)}]h/2$$

By equating these two energy expressions we obtain an equation that permits direct calculation of the weight of chain necessary to avoid any dynamic impulse. It is

$$mg = \frac{MV^2}{h[\sqrt{(L^2 + h^2)} - h]}$$

This will inevitably lead to a substantially heavier chain than that required to resist the design drag force. In fact, in this case, energy principles provide the only reliable guide to establishment of a design criterion. Corrections may be made for the effect of sag and chain elasticity. Such matters tend to represent fine tuning and, in this case, the net effect is small since an allowance for sag increases the required mg, while inclusion of elasticity reduces the required mg because some elastic strain energy can be stored.

1.3   Response to a suddenly applied load

In the first part of this section we deal with the theory of undamped response to a suddenly applied load of constant magnitude. This brief exposition is followed by a lengthy example.

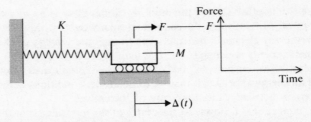

Figure 1.8 A suddenly applied force of constant magnitude F.

Figure 1.8 shows a definition diagram for this case. In solving the problem we shall use an energy approach.

The work done by a suddenly applied force of constant magnitude F is simply $F\Delta$, so the energy balance in the absence of damping is

$$\tfrac{1}{2}M\dot{\Delta}^2 + \tfrac{1}{2}K\Delta^2 = F\Delta \tag{1.11}$$

We shall assume that the oscillator starts from rest, that is, the initial conditions are $\Delta(0) = \dot{\Delta}(0) = 0$. Note from Equation 1.11 that $\dot{\Delta} = 0$ whenever $\Delta = 0$ or $\Delta = 2F/K$. This latter displacement is the peak displacement possible (true, by definition, because then $\dot{\Delta} = 0$), namely

$$\Delta_{max} = 2F/K \tag{1.12}$$

Furthermore, $\Delta = 0$ is the minimum value because the left-hand side of Equation 1.11 can never be negative (since a sum of squares is positive).

If the load had been gradually applied up to the full value F there would have been no dynamic response, only a static response given by

$$\Delta_{st} = F/K \tag{1.13}$$

By again utilising a plot of velocity versus displacement we now show that response to a suddenly applied load is again given by a circle whose centre is at the origin.

Let $\delta = \Delta - \Delta_{st}$. Hence $\dot{\delta} = \dot{\Delta}$. The energy balance becomes, after some rearrangement,

$$\tfrac{1}{2}M\dot{\delta}^2 + \tfrac{1}{2}K\delta^2 = \tfrac{1}{2}K\Delta_{st}^2$$

or

$$\left(\frac{\dot{\delta}}{\sqrt{(K/M)}\Delta_{st}}\right)^2 + \left(\frac{\delta}{\Delta_{st}}\right)^2 = 1 \tag{1.14}$$

which is the equation of a circle of radius unity.

14

Alternatively, Equation 1.14 may be written

$$\left(\frac{\dot{\Delta}}{\sqrt{(K/M)}\,\Delta_{st}}\right)^2 + \left(\frac{\Delta - \Delta_{st}}{\Delta_{st}}\right)^2 = 1 \qquad (1.15)$$

which may be plotted as in Figure 1.9.

Point A represents conditions at the start (that is, $\Delta = \dot{\Delta} = 0$), whereas point B represents response at time t. From the diagram we see that the angle swept out is ωt, where ω is the natural circular frequency. Hence, from geometry

$$OC = -1 \cos \omega t$$

That is,

$$\frac{\Delta}{\Delta_{st}} - 1 = -\cos \omega t \qquad \text{or} \qquad \frac{\Delta}{\Delta_{st}} = 1 - \cos \omega t$$

Similarly,

$$CB = 1 \sin \omega t \qquad \text{or} \qquad \frac{\dot{\Delta}}{\sqrt{(K/M)}\Delta_{st}} = \sin \omega t$$

In other words, the time history of response is

$$\Delta(t) = \frac{F}{K}(1 - \cos \omega t) \qquad (1.16)$$

and

$$\dot{\Delta}(t) = \frac{F\omega}{K} \sin \omega t \qquad (1.17)$$

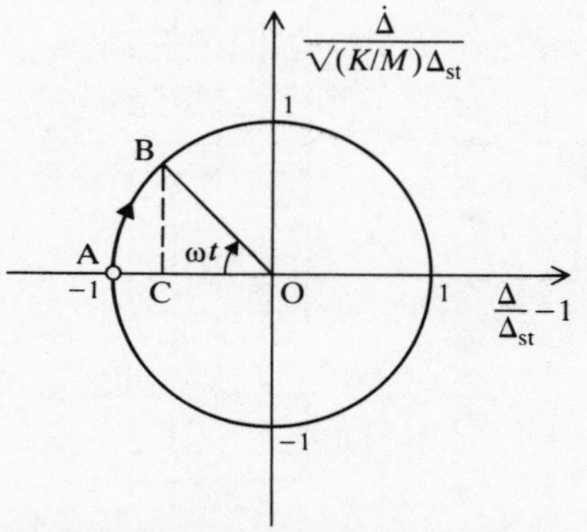

Figure 1.9 Response to a suddenly applied load as depicted in the velocity versus displacement plane.

15

Notice that the peak response is $2F/K$, a well known result in the annals of structural dynamics. This was, of course, predicted by Equation 1.12. As might be expected, inclusion of damping leads to concentric spiralling into the origin, resulting in a final deflection equal to the static deflection.

Example 1.3 Response of a 25-storey building to an idealised wind loading with a gust component

Earlier it was stated that the theory of the single-degree-of-freedom oscillator was a fertile ground for anyone interested in solving dynamics problems in practice. The present, albeit lengthy, example well exemplifies that statement. Its purpose is to illustrate a practical approach to the problem of the determination of the fundamental dynamic characteristics of a regular, if complicated, structural system.

A 25-storey by four-bay braced steel-frame building is shown in Figure 1.10. Individual frames are spaced at 6 m, bay size is constant at 6 m, and the interstorey height is also constant at 4 m. Apart from the top storey, the K braces in the central two bays carry over two floors.

In Figure 1.10 the full static design wind loading is shown idealised as a trapezoidal block varying in intensity from $5 \, \text{kN m}^{-1}$ at the base to $10 \, \text{kN m}^{-1}$ at the top. This corresponds to a steady wind of about $40 \, \text{m s}^{-1}$ at the top.

A preliminary design based on static wind loading is presented first. The braced core so 'designed' is then subjected to loading which is in part steady and in part sudden, in the form of a gust. Thus one arrives at an estimate of the dynamic amplification due to the gust loading.

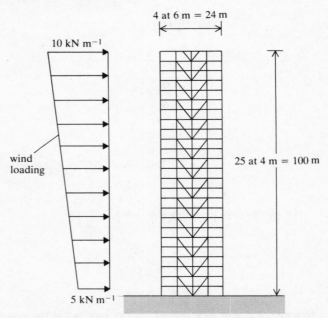

Figure 1.10 A 25-storey K braced frame.

STATIC RESPONSE

We consider first the static response of the frame to this wind loading. Lateral response is largely beam-like since the central two bays comprise a giant truss cantilevered from the foundation. The two chords of the truss (spaced at 12 m) resist the external bending moment due to the wind, and the shear force is resisted very efficiently by the K braces.

Setting aside the question of dead loading, suppose that the truss chords are proportioned such that the axial stress due to wind loading is constant with height − column section sizes being varied as necessary to effect this. That axial stress is

$$\sigma(z) = \frac{M_0(z)b/2}{I(z)} = \text{constant}$$

where b is the depth of the truss (12 m in this case), $M_0(z)$ is the overturning moment due to the wind at height z above the ground level, and $I(z)$ is the second moment of area of the column pair about the neutral axis. Thus

$$I(z) = A(z)b^2/2$$

in which $A(z)$ is the cross-sectional area of one chord at height z.

Let the lateral deflection of the frame be $\Delta(z)$, so the differential equation of flexure is

$$\frac{d^2\Delta}{dz^2} = \frac{M_0(z)}{EI(z)}$$

However, the right-hand side of above equation is constant and so

$$\frac{d^2\Delta}{dz^2} = \frac{2\sigma}{Eb} = \text{constant}$$

Integrating twice, we establish the curve of lateral deflection as the parabola

$$\Delta(z) = \frac{\sigma z^2}{Eb}$$

which satisfies the necessary conditions of zero deflection and zero slope at ground level (see Fig. 1.11).

Calling the peak deflection Δ_{des}, which occurs at the top of the building where $z = h$ (the height of the building is $h = 100$ m in this case), yields

$$\Delta_{des} = \left(\frac{\sigma h}{E}\right)\left(\frac{h}{b}\right)$$

The quantity $\sigma h/E$ is the total axial extension of the windward chord (say), while h/b is the aspect ratio of the frame. Hence the ratio of lateral deflection to axial deflection is equal to the aspect ratio, which is a tidy result.

17

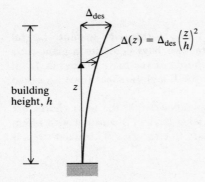

Figure 1.11 Calculated deflection profile for braced frame.

The deflected profile may be written as

$$\Delta(z) = \Delta_{des}(z/h)^2$$

and the only qualification is that the chords must have a finite size at the top − not zero as would be implied by the fact that the overturning moment is zero there. The discrepancy is insignificant since the deflection at the top is mostly influenced by stiffness at lower levels.

We have available two formulae. One, namely

$$\sigma = \frac{Eb}{h^2}\Delta_{des}$$

permits the axial stress in the chords to be checked, given a design for wind-load deflection. The second

$$A(z) = \frac{M_0(z)}{\sigma b}$$

then allows determination of the chord properties.

For instance, suppose we choose $\Delta_{des} = h/500 = 100/500 = 0.2$ m as the design tip deflection under full wind load. Since $h/b = 100/12 \approx 8$, and $E \approx 200\,000$ MPa, we have

$$\sigma = \frac{200\,000}{8} \times \frac{1}{500} = 50 \text{ MPa}$$

the modest size of which indicates that adequate allowance for dead-load stresses is possible. At the base of the frame the overturning moment of the wind is

$$M_0 = 5 \times 100 \times 100/2 + \tfrac{1}{2} \times 5 \times 100 \times \tfrac{2}{3} \times 100 = 4.17 \times 10^4 \text{ kN m}$$

Therefore, at the ground-floor level the truss chords must have an area

$$A = \frac{4.17 \times 10^4}{50 \times 10^3 \times 12} = 7 \times 10^{-2} \text{ m}^2$$

18

This requires a very substantial column section which may, in fact, need to be fabricated from individual plates: in that case a section with an overall depth of 450 mm, flange width of 400 mm, flange thickness of 70 mm and web thickness of 45 mm would suffice. Alternatively, a column core section CC 310 is almost adequate. Elsewhere section sizes would drop off − in four or five steps, say − as dictated by the overturning moment.

RESPONSE TO GUST LOADING

All the above is by way of a preamble to the main point of the exercise, which is to enquire what the situation would be if we consider part of the wind loading to be applied statically and part applied as a sudden gust loading.

Suppose, in fact, that half the wind loading is steady and the other half is applied suddenly. Half the full wind loading implies a steady wind of $40/\sqrt{2} \approx 28 \text{ m s}^{-1}$ at the top. Thus, the gust loading imposes a further 12 m s^{-1} on this steady wind. Gusts of this magnitude are possible although their duration would be short-lived, no more than a few seconds in length

The question that arises, therefore, is how response is affected by the different manner of application of the same total wind loading? There are two parts to the problem − steady loading and gust loading − and the obvious approach is to combine the effects of each. However, because the gust loading is short-lived, we need to obtain some quantitative basis for deciding the question of duration. So, once the steady loading component has been disposed of, we start work on writing the equation of motion for the building, thereby leading to the determination of the natural period of the structure. The length of the gust relative to the length of the fundamental period of the structure is the key to the problem.

The steady loading component is straightforward. The full static wind load caused a static deflection of 0.2 m, so half as much will give rise to a tip deflection at the top of 0.1 m. We may as well consider this the rest position prior to the arrival of the gust.

After the application of the gust loading the building acquires kinetic energy and the braced core stores more strain energy. The work done by the gust loading must equal the energy terms so

$$\text{KE} + \text{PE} = W$$

The increment in the top displacement, Δ_{dyn}, due to the gust, is the displacement quantity we shall work with. However, we need to know the curve of this additional (dynamic) deflection with height in order for the terms in the above energy balance to be evaluated. That is, we need to know $\Delta(z, t)$ ahead of time, which is tantamount to saying that we need to know $f(z)$ in the relationship

$$\Delta(z, t) = \Delta_{\text{dyn}}(t) f(z)$$

There is no reason to suppose that the dynamically deflected shape at some instant will be significantly different in form from the static response curve under the same loading − the magnitude will, of course, differ but the curves will look similar. In point of fact, advanced techniques of dynamic analysis show this to be an excellent

19

assumption and, furthermore, it allows us to model the structure as a single-degree-of-freedom system.

Therefore, we assume the deflected shape to be

$$\Delta(z, t) = \Delta_{\mathrm{dyn}}(t)(z/h)^2$$

where $\Delta_{\mathrm{dyn}}(t)$ is the time-varying increment in the tip displacement and $(z/h)^2$ is the shape function, determined earlier, that permits displacement levels to be established at various points up the building.

In general terms, if the magnitude of the gust loading is of intensity $p(z)$ per unit height, the work done by it is

$$W = \int_0^h p(z)\,\Delta(z, t)\,\mathrm{d}z$$

$$= \left(\int_0^h p(z)(z/h)^2\,\mathrm{d}z \right) \Delta_{\mathrm{dyn}}(t)$$

or

$$W = P_e\,\Delta_{\mathrm{dyn}}(t)$$

where

$$P_e = \int_0^h p(z)(z/h)^2\,\mathrm{d}z$$

and is defined as the equivalent lateral loading: it is always less than the actual applied loading

$$P = \int_0^h p(z)\,\mathrm{d}z$$

The vibrational velocity is

$$\frac{\partial}{\partial t}(\Delta(z, t)) = (z/h)^2\,\dot{\Delta}_{\mathrm{dyn}}$$

so the kinetic energy of vibration is

$$\mathrm{KE} = \tfrac{1}{2}\int_0^h m(z)\left(\frac{\partial\Delta}{\partial t}\right)^2\,\mathrm{d}z = \tfrac{1}{2}\left(\int_0^h m(z)(z/h)^4\,\mathrm{d}z \right) \dot{\Delta}^2_{\mathrm{dyn}}$$

where $m(z)$ is the mass per unit height of the building associated with one frame. The expression for the total kinetic energy may thus be written

$$\mathrm{KE} = \tfrac{1}{2}M_e\,\dot{\Delta}^2_{\mathrm{dyn}}$$

20

where

$$M_e = \int_0^h m(z)(z/h)^4 \, dz$$

and is the equivalent mass: it also is always less than the actual mass

$$M = \int_0^h m(z) \, dz$$

Finally, the elastic strain energy stored in the core is

$$PE = \tfrac{1}{2} \int_0^h EI(z) \left(\frac{\partial^2 \Delta}{\partial z^2} \right)^2 dz$$

$$= \tfrac{1}{2} \left(\frac{4}{h^4} \int_0^h EI(z) \, dz \right) \Delta_{dyn}^2$$

where $EI(z)$ has already been determined when the chords of the K brace truss were sized earlier. Clearly, it makes sense to define an equivalent lateral stiffness K_e such that

$$K_e = \frac{4}{h^4} \int_0^h EI(z) \, dz$$

and so we may write

$$PE = \tfrac{1}{2} K_e \Delta_{dyn}^2$$

Consequently, the energy balance takes on the familiar form

$$\tfrac{1}{2} M_e \dot{\Delta}_{dyn}^2 + \tfrac{1}{2} K_e \Delta_{dyn}^2 = P_e \Delta_{dyn}$$

and we have, now, explicit expressions for the correct mass, stiffness and loading terms that are associated with the chosen, representative, deflection. It follows immediately, incidentally, that we have an expression for the natural period of the system, namely

$$T = 2\pi \sqrt{(M_e/K_e)}$$

We shall calculate the period later. First, however, we establish the equation of motion or statement of dynamic equilibrium.

It will be recalled from Equation 1.2 that the rate at which energy is stored is equal to the rate at which work is done. That is

$$\frac{d}{dt} \left(\tfrac{1}{2} M_e \dot{\Delta}_{dyn}^2 \right) + \frac{d}{dt} \left(\tfrac{1}{2} K_e \Delta_{dyn}^2 \right) = \frac{d}{dt} \left(P_e \Delta_{dyn} \right)$$

21

Now, by the chain rule of differentiation

$$\frac{\mathrm{d}}{\mathrm{d}t}(\tfrac{1}{2}M_e\dot{\Delta}_{\mathrm{dyn}}^2) = \tfrac{1}{2}M_e\frac{\mathrm{d}}{\mathrm{d}\dot{\Delta}_{\mathrm{dyn}}}(\dot{\Delta}_{\mathrm{dyn}}^2)\frac{\mathrm{d}\dot{\Delta}_{\mathrm{dyn}}}{\mathrm{d}t}$$

$$= \dot{\Delta}_{\mathrm{dyn}}(M_e\ddot{\Delta}_{\mathrm{dyn}})$$

$$\frac{\mathrm{d}}{\mathrm{d}t}(\tfrac{1}{2}K_e\Delta_{\mathrm{dyn}}^2) = \tfrac{1}{2}K_e\frac{\mathrm{d}}{\mathrm{d}\Delta_{\mathrm{dyn}}}(\Delta_{\mathrm{dyn}}^2)\frac{\mathrm{d}\Delta_{\mathrm{dyn}}}{\mathrm{d}t}$$

$$= \dot{\Delta}_{\mathrm{dyn}}(K_e\Delta_{\mathrm{dyn}})$$

and

$$\frac{\mathrm{d}}{\mathrm{d}t}(P_e\Delta_{\mathrm{dyn}}) = \dot{\Delta}_{\mathrm{dyn}}(P_e)$$

The rate equation for energy becomes

$$\dot{\Delta}_{\mathrm{dyn}}(M_e\ddot{\Delta}_{\mathrm{dyn}} + K_e\Delta_{\mathrm{dyn}} - P_e) = 0$$

and since this statement must be valid for all times (that is, times at which $\dot{\Delta}_{\mathrm{dyn}}$ is non-zero as well as instants when it is zero) we have the equation of motion

$$M_e\ddot{\Delta}_{\mathrm{dyn}} + K_e\Delta_{\mathrm{dyn}} = P_e$$

This equation could have been written directly from Equation 1.3 – Newton's second law – but, as we have repeatedly stated, we would not have known what the equivalent mass, stiffness and loading terms were. It is also evident, again as was stated earlier, that Newton's law and the energy balance are intimately related.

From Equation 1.16 the solution that satisfies initial conditions of $\Delta_{\mathrm{dyn}}(0) = \dot{\Delta}_{\mathrm{dyn}}(0) = 0$ is

$$\Delta_{\mathrm{dyn}}(t) = \frac{P_e}{K_e}\left[1 - \cos\left(\frac{2\pi t}{T}\right)\right]$$

where $T = 2\pi\sqrt{(M_e/K_e)}$. Substitution in the equation of motion shows that the above solution satisfies it identically. Incidentally, because the applied load is constant and not variable or oscillatory, the only timescale possible in the response is that representative of the structure itself, namely, T. While this is true of free vibrational problems it is usually not the case when forcing is present.

In terms of the whole building the response is

$$\Delta(z, t) = \frac{P_e}{K_e}\left[1 - \cos\left(\frac{2\pi t}{T}\right)\right]\left(\frac{z}{h}\right)^2$$

Plotting the displacement response indicates that the peak dynamic displacement (namely $\Delta_{\max} = 2P_e/K_e$) first occurs at a time $t = T/2$ after the gust arrives (see Fig. 1.12a).

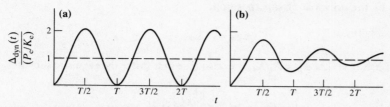

Figure 1.12 Time history of response to the gust loading: (a) undamped; (b) damped.

The inclusion of damping attenuates the response (as in Fig. 1.12b), and after some time we should be left with a result which is the same as if the gust loading was gradually applied. That is, we should be left with the static displacement due to $p(z)$: call this displacement Δ_{st}. Thus, we expect that, because of damping,

$$\lim_{t \to \infty} \Delta_{dyn}(t) \to \Delta_{st}$$

or, in other words,

$$\Delta_{st} = P_e/K_e$$

However, when one looks at the expressions for P_e and K_e it is not obvious that the static displacement due to $p(z)$, namely Δ_{st}, is given by the ratio P_e/K_e.

To lay the matter to rest it is necessary to digress and look a little deeper. Suppose the full static design wind loading is $\alpha p(z)$, where $p(z)$ is the gust loading and α is a constant greater than one. Recall from our earlier formulation that

$$P_e = \int_0^h p(z)(z/h)^2 \, dz$$

and

$$K_e = \frac{4}{h^4} \int_0^h EI(z) \, dz$$

However, because $EI(z)$ is determined from the design static loading of $\alpha p(z)$, we have

$$EI(z) = \frac{h^2}{2\Delta_{des}} M_0(z)$$

where $M_0(z)$ is the overturning moment due to $\alpha p(z)$ and Δ_{des} is the tip deflection under $\alpha p(z)$. As a result

$$K_e = \frac{2}{h^2 \Delta_{des}} \int_0^h M_0(z) \, dz$$

23

But, by the principle of superposition,

$$\Delta_{des} = \alpha \, \Delta_{st}$$

so, if it is true that $\Delta_{st} = P_e/K_e$, it is also true that (after the requisite substitutions)

$$\int_0^h \alpha p(z) z^2 \, dz = 2 \int_0^h M_0(z) \, dz$$

In other words, by proving the equality directly above, we prove that $\Delta_{st} = P_e/K_e$.

The simplest proof is that which invokes the theorem of virtual work. Consider a cantilever under load $\alpha p(z)$ which generates bending moment $M_0(z)$ (see Fig. 1.13). Consider the virtual displacement field $\Delta(z) = z^2$ with curvature $d^2\Delta/dz^2 = 2$. The external virtual work is

$$\int_0^h \alpha p(z) \Delta(z) \, dz = \int_0^h \alpha p(z) z^2 \, dz$$

The internal virtual work is

$$\int_0^h M_0(z) \frac{d^2\Delta}{dz^2} \, dz = 2 \int_0^h M_0(z) \, dz$$

and so, from the virtual-work theorem, we establish the result.

Incidentally, this result for the cantilever is completely general: one may adopt a similar procedure for, say, a simply supported beam, leading to the corresponding equilibrium theorem of

$$\int_0^h \alpha p(z) z(h - z) \, dz = \int_0^h M_0(z) \, dz$$

Sufficient theory has now been given. It remains to finish off the details with respect to our 25-storey building.

The period of the structure is found as follows. A reasonable value for the average specific weight of a steel-frame building (based on full dead load plus some live loading) is $1 \, kN \, m^{-3}$. Since individual frames are spaced at 6 m, and the whole bay width of 24 m is involved, the mass per unit height associated with one frame is

$$m = \frac{1 \times 6 \times 24}{9.81} = 14.7 \, kN \, s^2 \, m^{-2}$$

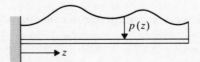

Figure 1.13 A cantilever under a general loading.

24

The equivalent mass is

$$M_e = \int_0^h m(z)(z/h)^4 \, dz = \frac{mh}{5} = \frac{14.7 \times 100}{5} \, \text{kN s}^2 \text{m}^{-1}$$

$$= 2.94 \times 10^2 \, \text{kN s}^2 \text{m}^{-1}$$

To get the equivalent stiffness we use a dodge. We have just shown that

$$\Delta_{st} = P_e/K_e$$

hence

$$K_e = P_e/\Delta_{st}$$

Here $\Delta_{st} = 0.1$ m (half the design wind loading gives rise to half the deflection, viz. $0.2/2 = 0.1$), and the equivalent lateral loading is

$$P_e = \int_0^h p(z)(z/h)^2 \, dz = \int_0^{100} (2.5 + 2.5z/100)(z/100)^2 \, dz$$

$$= 1.46 \times 10^2 \, \text{kN}$$

Consequently,

$$K_e = 1.46 \times 10^3 \, \text{kN m}^{-1}$$

and from the period formula

$$T = 2\pi \sqrt{\left(\frac{M_e}{K_e}\right)} = 2\pi \sqrt{\left(\frac{2.94 \times 10^2}{1.46 \times 10^3}\right)} = 2.8 \, \text{s}$$

The peak deflection comes after a time of $T/2$ or 1.4 s during which time it is unlikely that the gust has moved past the building. For most multistorey buildings (except the very tall ones or very slender structures), this approximation to gust action is equivalent to a suddenly applied load of constant magnitude because the peak deflection will have had time to develop.

Therefore, the peak displacement of the tip of the building is

$$\tfrac{1}{2}\Delta_{des} + 2\Delta_{st} = 0.1 + 0.2 = 0.3 \, \text{m}$$

representing a 50% increase over the design deflection of 0.2 m.

As has been amply demonstrated, a suddenly applied load gives rise to peak deflections that are twice as great as those resulting from the gradual application of the same loading. One would expect that associated quantities, such as shears and overturning moments, exhibit identical increases. We shall show this to be true but, as we shall presently see, the method of demonstration needs to be chosen carefully: at least one approach is inconclusive in this regard because we are dealing with a formulation that is approximately correct only.

25

Under the suddenly applied gust loading $p(z)$ the overturning moment is

$$M(z, t) = EI(z) \frac{\partial^2 \Delta(z, t)}{\partial z^2}$$

or

$$M(z, t) = \frac{2EI(z)}{h^2} \Delta_{\mathrm{dyn}}(t)$$

However, since $\Delta_{\mathrm{dyn}}(t)$ is at most twice the static deflection Δ_{st}, we have the expected result

$$M_{\max} = 2M_{\mathrm{st}}$$

where $M_{\max}$ is the peak bending moment due to the gust loading and M_{st} is the bending moment arising if the gust loading was gradually applied. The peak bending moment occurs at the same time as the peak displacement.

The shear force generated is

$$\frac{\partial}{\partial z}(M(z, t)) = \frac{\partial}{\partial z}(EI(z)) \frac{\partial^2 \Delta}{\partial z^2} + EI(z) \frac{\partial^3 \Delta}{\partial z^3}$$

$$= \frac{\partial}{\partial z}(EI(z)) \frac{\partial^2 \Delta}{\partial z^2}$$

because $\partial^3 \Delta / \partial z^3 = 0$ (Δ is proportional to z^2 in the present example). Again, making the necessary substitutions establishes the result that the peak dynamic shear force is twice the static value.

There is another way of proving that the peak base shear, for argument's sake, is twice the total lateral force of the gust loading. In the equivalent system the peak base shear is

$$S_{\mathrm{e}} = 2K_{\mathrm{e}}\Delta_{\mathrm{st}}$$

The true peak base shear is

$$S = 2K\Delta_{\mathrm{st}}$$

where K is the true stiffness of the braced frame.

Note that we have chosen the representative deflection of the equivalent system to be identical with the actual deflection of the true system. On this basis, the ratio P/K, where P is the true total lateral load acting, is exactly the same as the ratio $P_{\mathrm{e}}/K_{\mathrm{e}}$ for the equivalent system (remember that we used the virtual-work theorem, in effect, to prove this a little earlier). That is

$$P/K = P_{\mathrm{e}}/K_{\mathrm{e}}$$

26

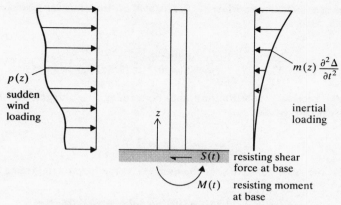

Figure 1.14   Free-body diagram for dynamic loading.

and so

$$S = 2P$$

The peak base shear is precisely twice the true static value.

All this is reassuring and it therefore appears that the model we have chosen is dynamically correct in all important aspects. However, this is not quite the case. To illustrate this, and thereby bring out the approximate nature of the assumptions made in formulating the problem, consider yet another way of establishing the base shear (see Fig. 1.14). It is

$$S(t) = \int_0^h p(z)\, dz - \int_0^h m(z)\, \frac{\partial^2 \Delta}{\partial t^2}(z, t)\, dz$$

The second term accounts for the inertia forces (product of mass and acceleration) due to the vibration: the first term represents the total gust loading, of course. Writing

$$S_{\mathrm{st}} = \int_0^h p(z)\, dz$$

gives

$$\frac{S(t)}{S_{\mathrm{st}}} = 1 - \ddot{\Delta}_{\mathrm{dyn}}(t)\, \frac{\int_0^h m(z)(z/h)^2\, dz}{\int_0^h p(z)\, dz}$$

But

$$\ddot{\Delta}_{\mathrm{dyn}}(t) = \frac{d^2}{dt^2}\left(\frac{P_{\mathrm{e}}}{K_{\mathrm{e}}}\left[1 - \cos\left(\frac{2\pi t}{T}\right)\right]\right) = \frac{P_{\mathrm{e}}}{M_{\mathrm{e}}}\cos\left(\frac{2\pi t}{T}\right)$$

27

and $S(t)$ is first a maximum when $t = T/2$, which is also when the displacement is a maximum. Hence

$$\frac{S_{max}}{S_{st}} = 1 + \frac{\int_0^h m(z)(z/h)^2 \, dz \int_0^h p(z)(z/h)^2 \, dz}{\int_0^h m(z)(z/h)^4 \, dz \int_0^h p(z) \, dz}$$

Making the necessary substitutions and performing the integrations for our 25-storey structure yields

$$S_{max}/S_{st} = 1 + \tfrac{5}{9} = 1.55$$

Similarly, the same approach for the moment at the base of the frame yields

$$\frac{M_{max}}{M_{st}} = 1 + \frac{\int_0^h m(z)(z/h)^3 \, dz \int_0^h p(z)(z/h)^2 \, dz}{\int_0^h m(z)(z/h)^4 \, dz \int_0^h p(z)(z/h) \, dz}$$

$$= 1 + \tfrac{21}{40} = 1.52$$

a more or less equivalent result. Clearly, in neither case is the number 2 recovered, yet the general approach is perfectly valid.

There is the whiff of an error here because we must have zero shear and zero moment at time $t = 0$ for the simple reason that the building has not flexed. Thus, the numbers $\frac{5}{9}$ and $\frac{21}{40}$ should both be unity in order that this be so. In fact, the answers are incorrect and the reason is that the true acceleration is not $\partial^2 \Delta(z, t)/\partial t^2$, as given above, but something larger than this. Our simplified model has not captured acceleration behaviour well, and this is due to the fact that accelerations are usually significantly influenced by higher-mode contributions whereas displacements, for example, are not. We have developed an equivalent single-degree-of-freedom model of a structure which has, in reality, an infinite number of degrees of freedom. For instance, the second mode of vibration of the building has a form similar to that shown in Figure 1.15b.

The natural period of that second mode is about one-tenth of the fundamental natural period. In other words, the natural circular frequency is an order of magnitude greater. Acceleration is proportional to the product of displacement and the square of the natural frequency, that is, $\ddot{\Delta} \sim \omega^2 \Delta$. Hence, even if the second-mode displacement was only 1% of that of the first mode, the acceleration in the second mode could be of the same order of magnitude as the acceleration in the first mode.

In point of fact the true acceleration at the top of the building is

$$\ddot{\Delta}_{dyn} \approx \frac{2}{(1 + \frac{5}{9})} \frac{d^2}{dt^2} \left(\Delta_{st} \left[1 - \cos \left(\frac{2\pi t}{T} \right) \right] \right)$$

The peak acceleration occurs at the instant the gust hits and is

$$\ddot{\Delta}_{max} \approx \frac{72\pi^2}{14T^2} \Delta_{st}$$

28

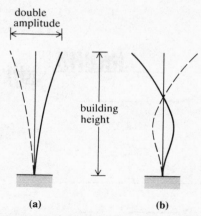

Figure 1.15 The first and second modes of vibration of the building: (a) first mode; (b) likely form of second mode.

Substituting in the relevant numbers for our gust loading gives

$$\ddot{\Delta}_{max} \approx \frac{72\pi^2}{14 \times 2.8^2} \times 0.1 \approx 0.65 \text{ m s}^{-2}$$

This is about 6.5% g and would certainly be quite noticeable to a person on the top floor of the building. Notwithstanding this, a gust of 12 m s^{-1} on top of 28 m s^{-1} is a rare event and may never occur in the life of the building.

A concluding comment is that the approach outlined in this example has, in part, similarities to the gust-factor method in wind engineering. See, for example, Davenport (1961, 1967) and Simiu and Scanlan (1977).

Example 1.4   Torsional lurch in a major suspension bridge on opening day

We use the final example in this chapter to expand on the statement given above, namely, that suddenly applied loads give rise to dynamic effects in structures that may be severely disquieting to people in or on them. A case in point concerns a major suspension bridge opened in the early 1970s. On opening day a huge crowd was on the bridge when a naval flotilla went underneath. People lined one side of the bridge and then moved quickly across to the other side as the ships passed underneath. The resulting torsional lurch caused many people to panic.

Figure 1.16 shows an idealised cross section of the bridge, with pedestrians representing loading on one side. The statically equivalent loads on the centreline are a vertical load p (representing the weight of people) and a torque m_t (representing the product of p and the distance to its centre of gravity). People moving across to the other side reverses the sign of m_t while p remains constant. If the bridge deck was under a slight twist θ due to m_t, then sudden application of $-2m_t$ (this is what is required to reverse the sign of m_t) would give rise to a twist of -4θ. The anti-clockwise lurch could have been as much as three times as great as the clockwise twist. What is more, the ensuing vibration gave rise to several peaks of appreciable magnitude before damping nullified it.

29

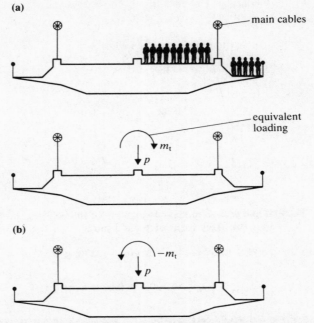

Figure 1.16 Pedestrian loads on opening day: (a) before; (b) after.

Fortunately there was no loss of life and subsequent checks of the structure indicated no deleterious effects. Nevertheless, the peak magnitude of the live loading was high and it came from a quite unexpected quarter.

1.4 Concluding remarks

In this chapter we have developed a little theory and supplemented this with several examples. In setting up problems it has frequently proved convenient to use the principle of conservation of energy. In some instances – where, for example, a complicated structural system is to be modelled as having a single degree of freedom – this is the only effective way of quickly being able to put numbers to desired quantities. In other instances an energy approach allows the simultaneous setting up and solving of a problem.

The energy approach to setting up equations of motion, as briefly presented here, can be made formal and elegant, but only at the risk of losing the simplicity of procedure that is so important for an introductory work such as this. As a result the principal object has been to sow the seeds of a methodology that should stand the user in good stead when attempting to set up a problem for subsequent dynamic analysis (see, for example,

Problem 1.7). The method is sometimes called Rayleigh's method after the famous 19th-century physicist who, through his celebrated work *The Theory of Sound*, did much to advance the mathematical and physical principles of dynamics.

Finally, one loose end remains to be tied up and that concerns the rate equation (Eqn 1.2). In the general case of a time-varying external force, the increment in work done is

$$dW = P(t)\, d\Delta$$

and so, by definition,

$$\frac{dW}{dt} = P(t)\frac{d\Delta}{dt} = P(t)\,\dot\Delta \tag{1.18}$$

Consequently, Equation 1.2 should be rewritten as

$$\frac{d}{dt}(KE + PE) = P(t)\,\dot\Delta \tag{1.19}$$

When dissipation is included, the rate of energy dissipation is similarly $D(t)\,\dot\Delta$, where $D(t)$ is the dissipative force, and it is implied that the quantity $D(t)\,\dot\Delta$ is always negative, regardless of the sign of $\dot\Delta$. This is true in most applications, although Example 2.2 provides a counter-example. The rate equation becomes

$$\frac{d}{dt}(KE + PE) = P(t)\,\dot\Delta + D(t)\,\dot\Delta \tag{1.20}$$

Problems

1.1 (a) With reference to Figures 1.4 and 1.5, obtain the peak lateral bending moment at the base of the crane boom due to a boom-tip slewing speed of $V_s = 1\ \mathrm{m\,s^{-1}}$. The payload is 200 kN and $L_1 = 20$ m, $L_2 = 40$ m.

(b) A crane is shown in idealised form in Figure 1.17. Points O, A, D are immovable, while OB is rigid but may rotate about O in response to stretching of the pendant AB. Find the equivalent stiffness of the cable system (in terms of EA_1, EA_2, L_1, L_2, and α and β) for vertical movement of the point C, for the special when AB is perpendicular to AO.

1.2 One truss of a two-lane steel truss bridge carries a dead load of $10\ \mathrm{kN\,m^{-1}}$, and the design live load is equivalent to $12.5\ \mathrm{kN\,m^{-1}}$ over the whole span. The overall dimensions of the truss are as shown in Figure 1.18.

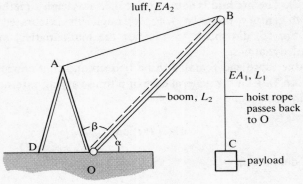

luff, EA_2

A

—boom, L_2

β

D

α

O

B

EA_1, L_1

hoist rope
passes back
to O

C

payload

Figure 1.17

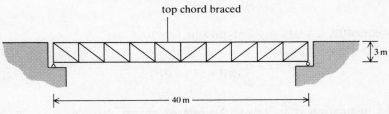

top chord braced

3 m

40 m

Figure 1.18

Assuming the chords are uniformly stressed along their length under dead load plus live load, show that the curve of deflection between the supports is parabolic.

If the permissible deflection under live load is 1/400 of the span, estimate the fundamental natural period of the bridge itself. Comment on the stresses in the chords of the truss under dead load plus live loading, given that the chords are 410 grade steel.

1.3 A tension-leg platform is anchored to the sea floor by four clusters of tubes, each cluster being under a tension T_0. If one cluster suddenly pulls away from its foundation, show that the peak tension recorded in each of the two closest clusters could be as high as $3T_0$.

1.4 A tugboat of mass m pulls a barge of mass M. If the hawser goes slack, after which the tug's velocity increases by δV, show that the length of hawser necessary to absorb adequately the jerk transmitted as the hawser tightens is given by

$$L = \frac{EA}{T_a^2} \frac{m(\delta V)^2}{(1 + m/M)}$$

where T_a is the allowable tension in the hawser.

32

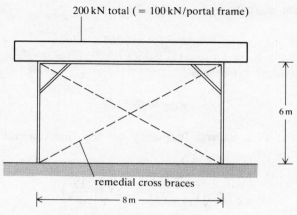

200 kN total (= 100 kN/portal frame)

6 m

remedial cross braces

8 m

Figure 1.19

1.5   A sand classifier (Fig. 1.19) in a quarry is highly susceptible to the movement of workers on the classifier platform. This is because its natural period in sway is estimated to be 0.9 s, and at this period it has proved possible for persons working on the platform inadvertently (and advertently!) to generate sizeable motions.

The structure in its present form is a simple portal frame with knee braces, and the full working load at platform level is 200 kN. Calculate the lateral stiffness of the existing structure, and estimate the diameter of the steel cross braces (proposed as remedial work) that will bring the period down to 0.3 s. With the structure so modified, it would lie outside the range in which it may be strongly excited by human movement.

Incidentally, if the original design was adequate for wind loading and the operation of the machinery on the platform had no definable periodicity, was the designer remiss in not applying a 'dynamic criterion of stiffness' to the problem?

1.6   The following data pertain to the mooring problem of Example 1.2: $L = 40$ m, $h = 20$ m, $V = 0.5$ m s^{-1}, $M = 800$ kN s^2 m^{-1}. Hence find the correct chain weight, translate this into a suitable link size, and check that the chain so chosen is adequate for the design drag force of 200 kN.

1.7   In the case of simple structures or structural elements such as beams, cantilevers, flat plates, etc., it is possible to find the fundamental natural frequency quite rapidly if the curve of deflection under self-weight is known or can be calculated.

For a uniform cantilever of length L, self-weight per unit length mg, and flexural rigidity EI, show that for a fundamental mode shape given by

$$\Delta(x, t) = \Delta_{\mathrm{dyn}}(t)[1 - \tfrac{4}{3}(x/L) + \tfrac{1}{3}(x/L)^4]$$

33

the equivalent mass is

$$M_e = 104 \, mL/405$$

and the equivalent stiffness is

$$K_e = 16 \, EI/5L^3$$

This gives rise to a natural frequency for the fundamental mode of a uniform cantilever

$$\omega \approx \sqrt{\left(\frac{162 \, EI}{13 \, mL^4}\right)} \approx 3.53\sqrt{\left(\frac{EI}{mL^4}\right)}$$

Likewise, in the case of a uniform simply supported beam show that for a mode shape given by

$$\Delta(x, t) = \Delta_{\text{dyn}}(t)\tfrac{16}{5}[(x/L) - 2(x/L)^3 + (x/L)^4]$$

$$M_e = 3968 \, mL/7875$$

and

$$K_e = 6144 \, EI/125 \, L^3$$

yielding

$$\omega \approx 9.88\sqrt{(EI/mL^4)}$$

The exact values are $3.53\sqrt{(EI/mL^4)}$ and $\pi^2\sqrt{(EI/mL^4)}$, respectively. In general, Rayleigh's method gives answers that are too high. That the agreement is excellent here indicates the power of Rayleigh's method when care is taken. The book by Blevins (1981) is a useful compendium of formulae for the natural frequencies and associated modes of vibration of a wide range of structural components.

1.8 A uniform beam of length L rests against a wall which is inclined at α to the horizontal. If the beam then slides onto a horizontal surface show that the velocity it acquires as it finally leaves the wall is given by

$$V = \sqrt{gL \sin \alpha}$$

Interpret this result for the case of a ship of length 100 m which is launched by skidding down a slipway inclined at $15°$ to the horizontal.

2

Theory of the single-degree-of-freedom oscillator

The aim of this chapter is to give a reasonably detailed account of the response of the single-degree-of-freedom oscillator to simple forcing functions. Free vibrations, including the effects of damping, are considered first. Section 2.2 deals with response to a few commonly used forcing functions. We also consider here the effect of system non-linearity in the form of elasto-plastic action. The emphasis is on obtaining peak response quantities. As before, several examples are given.

2.1 Free vibrations

2.1.1 The undamped system

The equation of motion for an undamped oscillator is

$$\ddot{\Delta} + \omega^2 \Delta = 0 \tag{2.1}$$

where $\omega = \sqrt{(K/M)}$ is the natural circular frequency of vibration (units of radians per second). The general solution is

$$\Delta(t) = A \cos \omega t + B \sin \omega t \tag{2.2}$$

and the specific solution for initial conditions of Δ_0 and $\dot{\Delta}_0$ is

$$\Delta(t) = \Delta_0 \cos \omega t + (\dot{\Delta}_0/\omega) \sin \omega t \tag{2.3}$$

If we define a constant ϕ such that $\sin \phi = \Delta_0/\sqrt{[\Delta_0^2 + (\dot{\Delta}_0/\omega)^2]}$ and $\cos \phi = (\dot{\Delta}_0/\omega)/\sqrt{[\Delta_0^2 + (\dot{\Delta}_0/\omega)^2]}$, Equation 2.3 may be written as

$$\Delta(t) = \sqrt{[\Delta_0^2 + (\dot{\Delta}_0/\omega)^2]}(\sin \phi \cos \omega t + \cos \phi \sin \omega t)$$

35

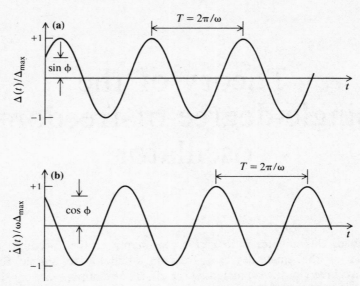

Figure 2.1 General solution for free undamped vibrations: (a) displacement response; (b) velocity response.

In view of the addition formula $\sin(\alpha + \beta) = \sin \alpha \cos \beta + \cos \alpha \sin \beta$, we have

$$\Delta(t) = \Delta_{max} \sin(\omega t + \phi) \qquad (2.4)$$

where $\Delta_{max} = \sqrt{[\Delta_0^2 + (\dot{\Delta}_0/\omega)^2]}$ is the amplitude of vibration (true since the peak values of $\sin(\omega t + \phi)$ are ± 1). The constant ϕ simply locates the start of the vibration relative to the origin and is referred to as the phase angle (see Fig. 2.1).

The following example centres on the calculation of peak displacement following an initial velocity imparted to the system.

Example 2.1 Airliner impact on a very tall building

Because of one or two near misses in the recent past, the owners of a very tall building have commissioned a study of the structural implications of the impact of a wide-bodied jet at the top of their building. Specifically, the owners (and their insurance companies!) wish to discover whether such an impact will cause more than serious local damage to the structure: they are particularly concerned to find whether the building will remain standing.

The aircraft is considered to be approaching a nearby runway–therefore its fuel load is low–and its weight and velocity are given as 2000 kN and 300 km h^{-1}, respectively. Figure 2.2 shows some of the gross details of the building and the purpose of the present exercise is to undertake a preliminary analysis.

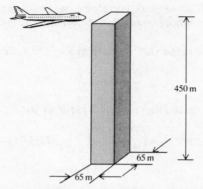

Figure 2.2 Definition diagram for aircraft impact on a tall building.

We shall consider the free vibrational response of the structure subject to a prescribed initial velocity. This permits calculation of the peak movement of the tip of the building after impact, which may be compared with the allowable deflection. Furthermore, it is thus possible to calculate an equivalent point load that gives rise to the same peak displacement.

On impact, all of the momentum of the aircraft is transferred to the building. For simplicity we shall assume that the resulting profile of the building deflection is linear with height. This is a reasonable compromise characterising the probable dual effects of flexure (giving a concave profile) and shear (leading to a convex profile). Hence, for a building of equal plan dimensions a, height b, and average specific weight γ, conservation of momentum yields

$$MV = \frac{\gamma a^2 b}{2g} v'$$

where M is the mass of the aircraft, V is its approach velocity and v' is the velocity with which the tip of the building begins to move. The factor $1/2$ arises because the average initial velocity of the building as a whole is $v'/2$, on account of the assumed linear deflection profile.

Thus, the initial velocity of the tip of the building is

$$\dot{\Delta}_0 = v' = \frac{2MVg}{\gamma a^2 b}$$

and from our previous work the free vibrational response is

$$\Delta(t) = (\dot{\Delta}_0/\omega) \sin \omega t$$

where $\omega = 2\pi/T = \sqrt{(K_e/M_e)}$, in which K_e is the equivalent building stiffness and M_e is the equivalent mass. The peak dynamic displacement is

$$\Delta_{\max} = \frac{\dot{\Delta}_0}{\omega} = \frac{MVgT}{\pi\gamma a^2 b}$$

37

In order to calculate Δ_{max} we need to evaluate the building period and we do this as follows. The design wind pressure for the building is given as the trapezoidal block shown in Figure 2.3.

Assuming a deflected shape that is linear and of the form

$$\Delta = \Delta_{des}(z/b)$$

the work done by this (gradually) applied wind load is

$$\tfrac{1}{2} \int_0^b ap(z)\Delta_{des}\left(\frac{z}{b}\right) dz = \tfrac{1}{2}\left(\frac{a}{b} \int_0^b p(z)z \, dz\right)\Delta_{des}$$

$$= \tfrac{1}{2} P_e \Delta_{des}$$

where Δ_{des} is the allowable wind drift at the top of the building and P_e is the equivalent wind load. The strain energy stored is

$$\tfrac{1}{2} K_e \Delta_{des}^2$$

where K_e is the equivalent lateral stiffness of the building. Equating these two expressions allows us to find K_e provided we set Δ_{des}. That is

$$K_e = P_e/\Delta_{des}$$

Now

$$P_e = \frac{65}{450} \int_0^{450} \left(1.5 + 1.5 \frac{z}{450}\right) z \, dz$$

$$= 3.66 \times 10^4 \text{ kN}$$

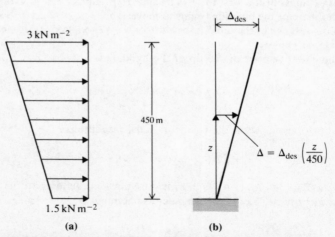

Figure 2.3 (a) Design wind pressure; (b) assumed deflection.

38

The design wind drift is believed to have been $\Delta_{des} = b/600 = 450/600 = 0.75$ m, so

$$K_e = 3.66 \times 10^4/0.75 = 4.88 \times 10^4 \text{ kN m}^{-1}$$

In keeping with the method outlined in Chapter 1, the equivalent mass for the vibrational study is

$$M_e = \int_0^b m(z)\left(\frac{z}{b}\right)^2 \mathrm{d}z = \frac{\gamma a^2 b}{3g}$$

The average specific weight of the building and its contents is estimated to be $\gamma = 1.25$ kN m^{-3} so the equivalent mass is

$$M_e = \frac{1.25 \times 65 \times 65 \times 450}{3 \times 9.81} = 8.08 \times 10^4 \text{ kN s}^2\text{ m}^{-1}$$

Therefore, the fundamental natural period of the structure is approximately

$$T = 2\pi\sqrt{(M_e/K_e)} = 2\pi\sqrt{(8.08/4.88)} = 8.1 \text{ s}$$

Consequently, the peak dynamic displacement is

$$\Delta_{max} = \frac{MVgT}{\pi\gamma a^2 b} = 2 \times 10^3 \left(\frac{300 \times 10^3}{60 \times 60}\right) \times \frac{8.1}{\pi \times 1.25 \times 65^2} \times \frac{1}{450}$$
$$= 0.18 \text{ m}$$

This is an appreciable deflection, being about 25% of the allowable wind drift, but it is safe to conclude that the overall structural integrity of the building will be maintained. Even so, local damage will be extremely severe and partial collapse of the uppermost floors is a distinct possibility as the aircraft engines will certainly demolish anything in their path.

In order to estimate the equivalent point force generated by the impact we could consider response to a suddenly applied load that gives Δ_{max}. If this load is P then the work done by its being suddenly applied is

$$P\Delta_{max}$$

The strain energy stored is

$$\tfrac{1}{2}K_e\Delta_{max}^2$$

so

$$P = \tfrac{1}{2}K_e\Delta_{max} = \tfrac{1}{2} \times 4.88 \times 10^4 \times 0.18 = 4.4 \times 10^3 \text{ kN}$$

This is more than twice the weight of the aircraft (however, see Example 2.4).

Table 2.1 gives a brief sensitivity analysis for varying wind-drift limitations and assumed deflected profiles.

Table 2.1 Sensitivity analysis for airplane impact problem.

	Linear deflection profile	Parabolic deflection profile
(a) Δ_{des}/height = 1/600		
T(s)	8.1	7.5
Δ_{max}(m)	0.18	0.25
P(kN)	4.4×10^3	4.1×10^3
(b) Δ_{des}/height = 1/450		
T(s)	9.3	8.6
Δ_{max}(m)	0.21	0.29
P(kN)	3.8×10^3	3.7×10^3

It may also be noted that, while momentum is conserved, energy is certainly not. The fraction of the airplane's energy lost on impact is

$$1 - \tfrac{1}{3}\left(\frac{\gamma a^2 b}{Mg}\right)\left(\frac{v'}{V}\right)^2 = 1 - \tfrac{4}{3}\left(\frac{Mg}{\gamma a^2 b}\right)$$

$$= 1 - \tfrac{4}{3} \times \frac{2 \times 10^3}{1.25 \times 65^2 \times 450}$$

$$= 99.9\%$$

Finally, and for the historical record: in 1945 a USAF bomber crashed into the 79th floor of the 102-storey Empire State Building in Manhattan. The crash occurred during the night and in fog. Building damage was local.

2.1.2 The damped system

Damping is present to some degree in all structural systems, but the nature of it and its magnitude are not well understood in any detailed sense. As a result of experimental work and full-scale tests, it is possible to assign damping levels to structural systems that bear some relation to reality. However, assignment of system damping still requires educated guesses to be made for the most part.

Internal friction at the microscopic level in the structural material and rubbing between structural and non-structural elements at the macroscopic level are two sources of damping. There is also the damping associated with the radiation of energy from a vibrating structure (e.g. through its foundations or through the surrounding water, as in an offshore structure). In such cases waves travel outwards from the source of disturbance never to return – hence the term 'radiation damping'. A further source is mobilised when the surrounding fluid – either air or water – is shifted and drag forces associated with system velocities are generated.

Even though several sources can be isolated it is common for damping to be taken either as viscous (i.e. the dissipation associated with moving a plunger backwards and forwards in oil) or purely frictional (i.e. Coulomb damping). In either case the damping coefficients are assumed constant, but there is experimental evidence to show that these damping coefficients are amplitude dependent (or, put another way, they increase as motion-generated stress levels increase).

In any event, the most common practice is to assume that viscous damping is present and to assign system damping levels based on experience and experimental evidence. There is, in fact, a widespread belief that if the gross nature of the dissipation is accounted for, the actual details are irrelevant for engineering calculations. That is, the concept of equivalent viscous damping holds sway.

We consider Coulomb damping first before concentrating on viscous damping, which will be the form assumed throughout the remainder of this book.

COULOMB DAMPING

In Coulomb damping one assumes the presence of a frictional force that is constant in magnitude but changes sign according to the sign of the vibrational velocity. In terms of energy rates we have (recall Eqn 1.20)

$$\frac{d}{dt}(KE + PE) = \dot{\Delta}(M\ddot{\Delta} + K\Delta) = f\dot{\Delta} \qquad (2.5)$$

where f is the constant frictional force.

The product $f\dot{\Delta}$ must remain negative irrespective of the periodically changing sign of the velocity $\dot{\Delta}$. Mathematically speaking f is to be regarded as positive when $\dot{\Delta}$ is negative, and vice versa. In a physical sense we say that f opposes $\dot{\Delta}$, as in Figure 2.4.

If an initial displacement Δ_0 is prescribed with zero initial velocity (so that after the start the velocity is at first negative, and we use $+f$), the equation of motion is

$$M\ddot{\Delta} + K\Delta = f \qquad (2.6)$$

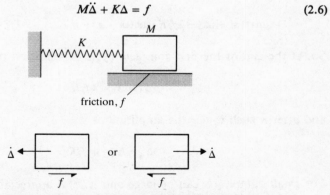

Figure 2.4 Schematic for Coulomb damping.

The solution may be written down (see Eqns 1.7 & 1.16) as

$$\Delta(t) = \Delta_0 \cos \omega t + \frac{f}{K}(1 - \cos \omega t)$$

where the first term on the right is the response to the initial conditions with zero forcing, and the second term on the right is the response to a suddenly applied load with zero initial conditions. By combining the two solutions we have invoked the principle of superposition. Rearranging the solution yields

$$\Delta(t) = (\Delta_0 - f/K)\cos \omega t + f/K \tag{2.7}$$

and the velocity is

$$\dot{\Delta}(t) = -\omega(\Delta_0 - f/K)\sin \omega t$$

which is negative until time $t = \pi/\omega = T/2$. This is, therefore, the limit on the applicability of the solution as portrayed above. Consequently, at the end of this first half cycle the peak displacement is

$$\Delta = -(\Delta_0 - 2f/K)$$

that is, the amplitude is reduced by an amount $2f/K$, which is itself the peak displacement due to a suddenly applied load f.

We do not need to consider the problem in any great detail beyond this point. The amplitude at the end of the next half cycle can be found by inspection. If at $t = T/2$ we prescribe an initial displacement of $-\Delta_0'$, where $\Delta_0' = \Delta_0 - 2f/K$, and we reverse the signs of each term in Eqn 2.7 we have

$$\Delta(t) = -(\Delta_0' - f/K)\cos(\omega t - \pi) - (f/K)[1 - \cos(\omega t - \pi)]$$

or

$$\Delta(t) = -(\Delta_0 - 3f/K)\cos(\omega t - \pi) - f/K \qquad T/2 \leqslant t \leqslant T \tag{2.8}$$

So, at the end of the first complete cycle of vibration the amplitude is

$$\Delta_1 = \Delta_0 - 4f/K$$

and after n such cycles the amplitude is

$$\Delta_n = \Delta_0 - 4nf/K$$

For small damping n can be large and Δ_n still appreciable, but eventually a stage is reached when the spring force cannot overcome the friction and

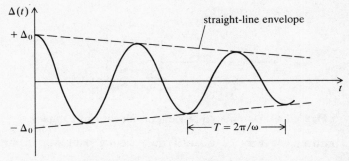

Figure 2.5 Decay of free vibration under the assumption of Coulomb damping.

vibration ceases. This occurs when $f = K\Delta_N$ and so

$$\Delta_N = \frac{f}{K} = \Delta_0 - \frac{4Nf}{K}$$

But, for all practical purposes, vibration ceases after N cycles where N is given by the nearest integer to the number $K\Delta_0/4f$.

Typical results are plotted in Figure 2.5. Note that the envelopes are given by two straight lines, each half-cycle is a cosine function, and the natural period of vibration is unchanged by this form of damping.

VISCOUS DAMPING

Figure 2.6 shows a representation of viscous damping action. Again, in terms of a rate equation we have

$$\frac{\mathrm{d}}{\mathrm{d}t}(\tfrac{1}{2}M\dot{\Delta}^2 + \tfrac{1}{2}K\Delta^2) = \dot{\Delta}(M\ddot{\Delta} + K\Delta) = D\dot{\Delta} \qquad (2.9)$$

where D is the damping force. For viscous damping it is assumed that $D = -c\dot{\Delta}$. That is, the damping force is directly proportional to the relative velocity between mass and support. The damping coefficient c is positive, with the result that the term $c\dot{\Delta}^2$ is always positive, regardless of the sign of $\dot{\Delta}$. Hence, the equation of motion is

$$M\ddot{\Delta} + c\dot{\Delta} + K\Delta = 0$$

which may be rearranged to

$$\ddot{\Delta} + 2\zeta\omega\dot{\Delta} + \omega^2\Delta = 0 \qquad (2.10)$$

where ω is the natural frequency in the absence of damping and $\zeta = c/2M\omega = c/2\sqrt{(KM)} = c/c_c$, where c_c is defined as the critical damping coefficient. The quantity ζ is thus defined as the fraction of critical damping

43

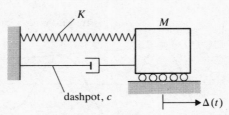

Figure 2.6 Dashpot representation for viscous damping.

and is a convenient way of quantifying viscous damping in structural systems.

The general solution to Equation 2.10 depends on the value of ζ. If $\zeta > 1$ the system is described as overdamped – the solution involves exponential functions, which are not periodic, and therefore this case is of somewhat limited interest (an important exception occurs in the case of pile driving, and Example 2.3 is therefore devoted to an exposition of this topic). Similarly, little interest attaches to the case when $\zeta = 1$ and the system is critically damped. In both, the character of the solution is akin to the motion of a door to which is attached a proprietary combination spring and heavy damping device – after being opened, the door slowly shuts.

However, by far the greatest interest attaches to the case when $\zeta < 1$, which is somewhat unfortunately referred to as underdamped. The general solution involves exponential functions and also circular trigonometric functions, and is therefore periodic. It reads

$$\Delta(t) = e^{-\zeta\omega t}(A \cos \omega_{\mathrm{d}}t + B \sin \omega_{\mathrm{d}}t)$$

where $\omega_{\mathrm{d}} = \sqrt{(1 - \zeta^2)}\omega$ is the natural circular frequency in the presence of viscous damping, and A and B are constants to be assigned for given initial conditions.

In particular, for initial conditions of Δ_0 and $\dot{\Delta}_0$ the solution is

$$\Delta(t) = e^{-\zeta\omega t}\left(\Delta_0 \cos \omega_{\mathrm{d}}t + \frac{\dot{\Delta}_0 + \zeta\omega\Delta_0}{\omega_{\mathrm{d}}} \sin \omega_{\mathrm{d}}t\right) \qquad (2.11)$$

which may be written more compactly as

$$\Delta(t) = \Delta_{\max}e^{-\zeta\omega t} \sin(\omega_{\mathrm{d}}t + \phi) \qquad (2.12)$$

where the amplitude is

$$\Delta_{\max} = \Delta_0 \sqrt{\left[1 + \left(\frac{\dot{\Delta}_0}{\omega_{\mathrm{d}} \Delta_0} + \frac{\zeta\omega}{\omega_{\mathrm{d}}}\right)^2\right]}$$

and the phase is

$$\phi = \tan^{-1}[\omega_{\mathrm{d}} \Delta_0/(\dot{\Delta}_0 + \zeta\omega\Delta_0)]$$

44

Clearly, the exponential function attenuates the magnitude of successive peaks in the vibration. Added to which, the rate of decay is not constant, as it was with Coulomb damping; rather, it decreases with time as the vibration velocity reduces. The general trend is illustrated in Figure 2.7.

One of the ways of estimating the damping level in a structure is to conduct a free vibration test and inspect the record or time history of response recorded. The theory is as follows. A positive peak in displacement occurs when t is such that $\sin(\omega_d t + \phi) = 1$ (see Eqn 2.12). It next occurs when $\sin(\omega_d t + \phi + 2\pi) = 1$, and so on. If δ is the ratio of two peaks n cycles apart, then

$$\delta = \frac{\exp[-\zeta\omega(t + nT_d)]}{\exp(-\zeta\omega t)} = \exp(-\zeta\omega n T_d) = \exp[-2n\pi\zeta/\sqrt{(1 - \zeta^2)}]$$

Taking natural logarithms of both sides yields

$$\frac{\zeta}{\sqrt{(1 - \zeta^2)}} = \frac{1}{2n\pi} \ln\left(\frac{1}{\delta}\right)$$

and the fraction of critical damping is seen to be dependent only on the amplitude ratio after n cycles have passed. In practice n needs to be of the order of 10 to ensure reasonable accuracy in reading δ. For small to moderate values of ζ (say $\zeta < 0.1$), the term $\sqrt{(1 - \zeta^2)}$ is insignificantly different from unity and we may thus, in general, ignore the effect of damping on the natural period and write the above equation as

$$\zeta = \frac{1}{2n\pi} \ln\left(\frac{1}{\delta}\right) \tag{2.13}$$

The quantity $\ln(1/\delta)$ is referred to as the logarithmic decrement. For example, if after 10 cycles of vibration the peak displacement has reduced by 50%, the fraction of critical damping indicated is

$$\zeta = \frac{1}{20\pi} \ln 2 = 0.011 \approx 1\%$$

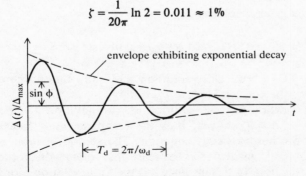

Figure 2.7 Decay of free vibration under the assumption of viscous damping.

45

Table 2.2 Suggested damping values when working stresses are not exceeded.

(1)	welded steel, prestressed concrete, well detailed reinforced concrete	$\zeta = 0.02$ to 0.03
(2)	reinforced concrete in which considerable cracking is apparent	$\zeta = 0.03$ to 0.05
(3)	bolted and/or riveted steel, wood structures with nailed or bolted joints	$\zeta = 0.05$ to 0.07

Clearly, a damping level of the order of 1% is by no means insignificant.

When damping is small there is only a small change in amplitude from one cycle to the next, and so the equally spaced concentric spirals sketched in Figure 1.3 are an accurate reflection of the decay of vibration in the velocity–displacement plane. It may also be noted that, since energy is proportional to the square of either velocity or displacement, the fractional energy loss per cycle is roughly twice the fractional loss in amplitude.

Damping values for structures lie typically in the range $0.01 < \zeta < 0.1$, with most structures exhibiting values in the central region, say from 0.02 to 0.05. Table 2.2, prepared by Newmark and Hall (1982), indicates values of ζ that have been recommended for situations in which working stress levels are not exceeded.

2.2 Response to simple forcing functions

2.2.1 Suddenly applied load of constant magnitude and infinite duration

Labelling such a force F, the equation of motion reads

$$\ddot{\Delta} + 2\zeta\omega\dot{\Delta} + \omega^2\Delta = F/M = \omega^2 F/K$$

for which the general solution is

$$\Delta(t) = e^{-\zeta\omega t}(A \cos \omega_d t + B \sin \omega_d t) + F/K \qquad (2.14)$$

where $\omega_d = \sqrt{(1 - \zeta^2)}\,\omega$. Differentiation and substitution shows that the solution is of the correct form.

The first portion of the solution is the solution to the equation $\ddot{\Delta} + 2\zeta\omega\dot{\Delta} + \omega^2 \Delta = 0$, that is, it is the complementary function or the solution to the homogeneous equation. The second portion adds in the particular integral, particular in the sense that it depends on the nature of the forcing function. In this instance it is the static solution. For initial conditions of $\Delta(0) = \dot{\Delta}(0) = 0$, the solution is

$$\Delta(t) = \frac{F}{K} \left[1 - e^{-\zeta \omega t} \left(\cos \omega_d t + \frac{\zeta}{\sqrt{(1 - \zeta^2)}} \sin \omega_d t \right) \right] \qquad (2.15)$$

where F/K is the static deflection, by definition. Notice that the solution, as portrayed by Equation 2.15, is linear in that response is doubled when the external loading is doubled, for example. This linearity is the justification for superposing particular integral and complementary function.

The peak response occurs at time $t \approx T_d/2$ and is

$$\Delta_{max} \approx \frac{F}{K} (1 + e^{-\zeta \pi}) \qquad (2.16)$$

When $\zeta = 0$ we recover the classical result of $\Delta_{max} = 2F/K$ or, put another way, the peak force to which the system is subjected is twice the applied force. When $\zeta = 0.1$, $\Delta_{max} = 1.73 \, F/K$ and, since this represents a high level of damping for most situations encountered in practice, we conclude that the effect of damping in this instance is negligible. This reinforces a point made earlier that damping needs time to manifest itself: in the present case the peak response occurs too early for any appreciable reduction in the peak.

The damping envelope is $e^{-\zeta \omega t}$ and it is clear that transients are damped out more quickly when the natural frequency is high than when the system has a low natural frequency. For example, similar terms describe the attenuation of the waves transmitted through the ground from the source region of an earthquake. In this instance time may be measured as distance from source divided by wave velocity, so that high-frequency components in the ground motion soon die away. However, distant earthquakes can set tall buildings, and other structures of relatively long period, swaying, but be otherwise imperceptible to people occupying structures whose natural periods are shorter.

Example 2.2   A system exhibiting negative damping

In the vast majority of situations damping is dissipative, that is, energy is lost by the vibrating system. However, there are important exceptions.

Negative damping, as the name implies, gives rise to vibrations that continually increase in magnitude. That is, energy is extracted from, rather than dissipated into, the surroundings. This type of vibration may be classed as self-excited: it includes aero-elastic phenomena such as flutter of wings and suspension bridges and the galloping of ice-laden wires.

A simple model of this phenomenon has been provided by Inglis (1963). It is also a useful way of linking the concepts of viscous damping and response to a suddenly applied load which has its origins in Coulomb-type friction. The argument is as follows.

Dry friction is generally found to be greatest prior to movement between two surfaces and to decrease as the relative velocity increases; a useful analogy is that of push starting a car. Thus a model of this type of friction which incorporates both Coulomb damping and viscous damping is for a friction force given by

$$f - cV_r$$

where V_r is the relative velocity between the surfaces.

Now consider a spring–mass system in which the mass sits on a horizontal conveyor belt and the spring is anchored independently. The situation is as shown in Figure 2.8. When the conveyor starts up, say with velocity V, the mass moves with the belt until the spring force overcomes the friction f between the touching surfaces. At this juncture the mass begins to vibrate because, in slipping, the friction force is reduced and equilibrium between it and the spring force can only be maintained by an inertia force being generated. Shortly afterwards the level of vibration is violent.

The problem may be separated into two parts. Until slippage first occurs the displacement of the mass simply follows the conveyor motion, that is

$$\Delta(t) = Vt$$

and slippage occurs when

$$K\Delta = f$$

which is at a time t_*, after the start, given by

$$t_* = f/KV$$

Subsequently, we have the equation of motion

$$M\ddot{\Delta} + K\Delta = f - cV_r$$

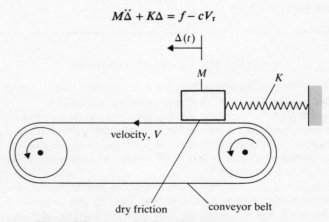

Figure 2.8 A model for self-excited vibrations.

48

where the relative velocity is $V_r = V - \dot\Delta$, the negative sign being correct since $\dot\Delta = V$ at $t = t_*$, and so it is still just true that $V_r = 0$. Just after time t_* slippage occurs and $\dot\Delta$ is less than V, so V_r is positive and f *is* reduced by the amount cV_r. Hence, we have

$$M\ddot\Delta - c\dot\Delta + K\Delta = f - cV$$

The right-hand side is equivalent to a suddenly applied load of constant magnitude, and negative viscous damping is clearly indicated on the left-hand side. Note that the Coulomb friction force f does not change sign as was the case with Equation 2.6 because in the present case the conveyor moves with a constant velocity.

In a formal sense we have now to solve

$$\ddot\Delta - 2\zeta\omega\dot\Delta + \omega^2\Delta = \omega^2\left(\frac{f - cV}{K}\right)$$

for $t \geqslant t_*$. The initial conditions are $\Delta(t_*) = f/K$, $\dot\Delta(t_*) = V$.

In constructing a solution we can use the solutions given by Equations 2.11 (free vibrations for given initial conditions) and 2.15 (forced vibration for zero initial conditions), adapt them, and then simply add them together (thereby once again invoking the principle of superposition). That is

$$\Delta(t) = \exp[\zeta\omega(t - t_*)]\left[\frac{f}{K}\cos[\omega_d(t - t_*)] + \left(\frac{V - \zeta\omega f/K}{\omega_d}\right)\sin[\omega_d(t - t_*)]\right]$$

$$+ \frac{f - cV}{K}\left[1 - \exp[\zeta\omega(t - t_*)]\left(\cos[\omega_d(t - t_*)] - \frac{\zeta}{\sqrt{(1 - \zeta^2)}}\sin[\omega_d(t - t_*)]\right)\right]$$

for $t \geqslant t_*$.

This may be rearranged into the relatively simple form

$$\Delta(t) = \frac{f - cV}{K} + \frac{V}{\omega_d}\exp[\zeta\omega(t - t_*)]\sin[\omega_d(t - t_*) + \phi]$$

where $\omega_d = \omega\sqrt{(1 - \zeta^2)}$ and $\tan\phi = 2\zeta\sqrt{(1 - \zeta^2)}/(1 - 2\zeta^2)$.

The solution as depicted above consists of a steadily growing sinusoidal response about a mean displacement line given by $(f - cV)/K$. This is shown in Figure 2.9. Note that the growth of vibration is most rapid for systems with high natural frequencies. In a mathematical sense, therefore, the above model is instructive.

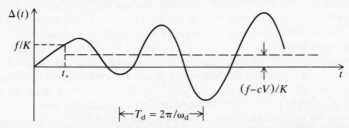

Figure 2.9 Time history of response for a system exhibiting negative damping.

49

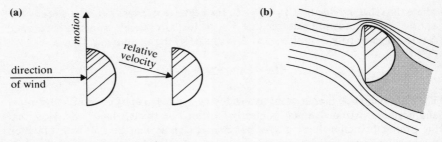

Figure 2.10 Schematic for a galloping ice-laden wire: (a) assumed cross section; (b) resultant air flow.

In a physical sense an excellent example of a self-excited oscillation is that of the galloping of an ice-laden wire. A direct quotation from Inglis (1963) is in order:

This phenomenon is exhibited by long-span power lines which, under the action of wind, may develop large vertical oscillations persisting for long periods of time. The action is associated with wintry condition when, owing to a deposit of ice, the wire loses its circularity of section. The explanation in broad outline is quite simple.

Let us take an extreme case and assume the section of the wire to be semi-circular with the flat side facing the wind [see Fig. 2.10a]. Imagine that the wind is blowing horizontally and that the wire is momentarily moving upwards. The velocity of the wind relative to the wire is then downwards, and the air-flow round it is as indicated [in Fig. 2.10b].

Around the top the air follows the surface, the lines of flow are crowded together, the velocity of flow is increased and consequently the air pressure is reduced. The lower part of the wire finds itself in a stagnant backwater where the pressure is approximately atmospheric. Thus, when moving upwards, the wire receives an upward urge and when moving downwards it will be acted upon by a downward force. Hence, even in a moderate breeze, a violent and persistent state of oscillation can be excited in a wire which has become non-circular in section.

Example 2.3   Pile driving

This lengthy example contains much that is of direct practical use and it is also the only example known to the author whereby viscous damping can be established from first principles, rather than merely assumed to exist. Furthermore, the level of damping can be determined exactly rather than guessed. It is, interestingly, one of the few systems which can in the normal course of events be overdamped.

We consider a system in which the hammer, or ram, and the capblock form a mass–spring system, leaving the pile to act as a wave guide radiating energy away from its upper face: the pile is the dashpot, in other words. Once again we are able to exploit the simple solutions established earlier, although the business of setting up the equations of motion is more involved in this instance. For this we can adapt the treatment by Irvine (1981) relating to tensile stress wave propagation in cables,

50

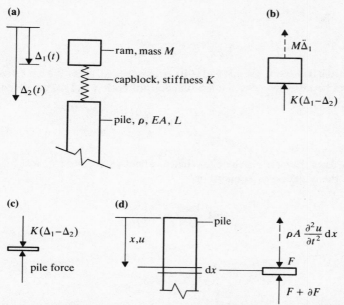

Figure 2.11 (a) Definition diagram for pile-driving equations. (b) Equilibrium of ram. (c) Equilibrium at capblock-pile interface. (d) Definition diagrams for pile force.

with the obvious difference being that at least the initial wave in the pile is compressive. In this treatment, reflections will not be considered. However, elastoplastic behaviour at the pile head is considered later in this chapter in Example 2.8.

The definition diagram shown in Figure 2.11 gives the details.

The equation governing the deceleration of the ram of mass M is

$$M\ddot{\Delta}_1 + K(\Delta_1 - \Delta_2) = 0$$

where Δ_1 is the absolute displacement of the ram after contact is made with the capblock, which is of compressive stiffness K. The displacement Δ_2 is the absolute displacement of the upper end of the pile at its junction with the capblock. Note that the weight of the ram has been left out of the above equation because it is generally insignificant when compared to the inertia force $M\ddot{\Delta}_1$ – after all the ram falls through a significant height and is stopped suddenly.

The other equation is a force balance between the capblock and the pile head. In order to determine the force in the pile head we need to do an analysis of the pile itself (see Fig. 2.11d).

Let x measure distance from the pile head, and let $u(x, t)$ be the axial displacement of the pile. Equilibrium of an element of the pile, of length dx, is

$$F + \partial F - F + \rho A \frac{\partial^2 u}{\partial t^2} \, dx = 0$$

51

or

$$\frac{\partial F}{\partial x} + \rho A \frac{\partial^2 u}{\partial t^2} = 0$$

where ρ is the density of the pile and A its cross-sectional area. We have used partial derivatives because both F and u are functions of both x and t. From Hooke's law

$$F = -EA \frac{\partial u}{\partial x}$$

where $\partial u/\partial x$ is the strain in the pile. Hence, when EA is constant, which is usually the case, the equation of equilibrium is

$$-EA \frac{\partial^2 u}{\partial x^2} + \rho A \frac{\partial^2 u}{\partial t^2} = 0$$

or

$$\rho \frac{\partial^2 u}{\partial t^2} - E \frac{\partial^2 u}{\partial x^2} = 0$$

A solution of this equation for a disturbance that travels down the pile is

$$u = f(x - V_d t) = f(X)$$

where f is a function determined by the initial conditions. This is D'Alembert's solution. It is assumed that V_d, the velocity with which information passes along the pile, is constant and this fact, together with an expression for V_d, can be found by substituting in the governing differential equation. Now

$$\frac{\partial u}{\partial t} = \frac{\mathrm{d}f}{\mathrm{d}X} \frac{\partial X}{\partial t} = -V_d f'$$

so

$$\frac{\partial^2 u}{\partial t^2} = V_d^2 f''$$

Also

$$\frac{\partial u}{\partial x} = \frac{\mathrm{d}f}{\mathrm{d}X} \frac{\partial X}{\partial x} = f'$$

so

$$\frac{\partial^2 u}{\partial x^2} = f''$$

Hence

$$(\rho V_d^2 - E)f'' = 0$$

In general $f'' \neq 0$, so

$$V_d = (E/\rho)^{1/2}$$

which is constant and is referred to as the dilatational wave speed. The time (L/V_d) is the time taken for a blow delivered at one end of the pile to reach the far end (note that, for steel, $V_d \sim 5000 \text{ ms}^{-1}$, the speed of sound in steel).

The force at some point in the pile is

$$F = -EA \frac{\partial u}{\partial x}$$

But from our earlier work we have the useful and important result that

$$\frac{\partial u}{\partial x} = -\frac{1}{V_d} \frac{\partial u}{\partial t}$$

so an alternative expression for the force in the pile is

$$F(x, t) = \frac{EA}{V_d} \frac{\partial u}{\partial t} = \rho A V_d \frac{\partial u}{\partial t}$$

Therefore, at the pile head, where $u \equiv \Delta_2$, we have the result

$$F(0, t) = \rho A V_d \dot{\Delta}_2$$

Thus, the force balance at the capblock–pile interface is

$$K(\Delta_1 - \Delta_2) = \rho A V_d \dot{\Delta}_2$$

Linking the two equations we obtain

$$M\ddot{\Delta}_1 = -\rho A V_d \dot{\Delta}_2$$

which may be integrated directly to

$$M\dot{\Delta}_1 = -\rho A V_d \Delta_2 + \text{constant}$$

The constant may be evaluated once we have established the initial conditions. The height through which the ram falls allows its striking velocity to be calculated, say, $\dot{\Delta}_0$. Thus

$$\dot{\Delta}_1(0) = \dot{\Delta}_0$$

and, of course, the pile head has no initial displacement (i.e. $\Delta_2(0) = 0$), so

$$M\dot{\Delta}_1 = M\dot{\Delta}_0 - \rho A V_d \Delta_2$$

Substituting for Δ_2 then gives, as the equation of motion for the ram–capblock system,

$$M\ddot{\Delta}_1 + \frac{KM}{\rho A V_d}\dot{\Delta}_1 + K\Delta_1 = \frac{KM}{\rho A V_d}\dot{\Delta}_0$$

with initial conditions of $\dot{\Delta}_1(0) = \dot{\Delta}_0$ and $\Delta_1(0) = 0$. This equation exhibits pure viscous damping (due to the radiation of energy away from the capblock by the pile) – which, as was mentioned earlier, can be quantified. Written in standard form the equation is

$$\ddot{\Delta}_1 + 2\zeta\omega\dot{\Delta}_1 + \omega^2\Delta_1 = 2\zeta\omega\dot{\Delta}_0$$

where $\zeta = \sqrt{(KM)}/2\rho A V_d$. The forcing term is constant, so we are able to exploit our earlier solutions.

UNDERDAMPED

For the case where $\zeta < 1$ the solution may be extracted from Equations 2.11 and 2.15 – comprising the sum of response to an initial velocity and response to a suddenly applied load. It reads

$$\Delta_1(t) = \frac{2\zeta\dot{\Delta}_0}{\omega}\left[1 - e^{-\zeta\omega t}\left(\cos\omega_d t + \frac{\zeta}{\sqrt{(1-\zeta^2)}}\sin\omega_d t\right)\right] + \frac{\dot{\Delta}_0}{\omega_d}e^{-\zeta\omega t}\sin\omega_d t$$

The force in the end of the pile is

$$F_p = \rho A V_d \dot{\Delta}_2 = -M\ddot{\Delta}_1$$

and after some rearrangement this can be shown to be

$$F_p = \frac{M\omega\dot{\Delta}_0}{\sqrt{(1-\zeta^2)}}e^{-\zeta\omega t}\sin\omega_d t$$

or

$$F_p = K\left(\frac{\dot{\Delta}_0}{\omega_d}e^{-\zeta\omega t}\sin\omega_d t\right)$$

By writing the equation in the form of capblock stiffness times dynamic displacement we are able to isolate an interesting fact: the displacement term in parentheses is that due to an initial velocity. In retrospect the ram–capblock–pile system could be viewed as in Figure 2.12. The pile force then is the reaction to the transmitted capblock force when the ram impinges on it with an initial velocity. Such a view can only be described as being blessed by much hindsight!

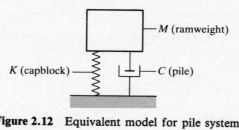

Figure 2.12 Equivalent model for pile systems.

Finally, the displacement of the pile head is found to be

$$\Delta_2(t) = \frac{2\zeta\dot{\Delta}_0}{\omega}\left[1 - e^{-\zeta\omega t}\left(\cos \omega_d t + \frac{\zeta}{\sqrt{(1-\zeta^2)}}\sin \omega_d t\right)\right]$$

which is the response to a suddenly applied load. It is not easy to describe this behaviour with a simple isolated model: nonetheless, the way the various solutions have separated is of some interest in itself.

Incidentally, peak values of pile force and pile displacement are

$$F_{p,\max} = K\dot{\Delta}_0/\omega_d, \qquad \Delta_{2,\max} = 4\zeta\dot{\Delta}_0/\omega.$$

for $\zeta < 1$. However, both tend to be excessively conservative and it is better to proceed on a case-by-case basis. A matter of considerable importance is the time of validity of the solution. The stress wave will reflect from the far end of the pile and travel back to the pile head. The elapsed time before this occurs is

$$2L/V_d$$

where L is the length of the pile. The solution we have is valid only for

$$t < 2L/V_d$$

Furthermore, the capblock cannot supply tension between ram and pile head. But the solution for the pile force shows that it could become tensile after time

$$\omega_d t = \pi$$

or

$$t = \pi/[\omega\sqrt{(1-\zeta^2)}]$$

So, likewise, the solution is valid only if, in addition,

$$t < \pi/[\omega\sqrt{(1-\zeta^2)}]$$

One or other of these limits governs the solution in its present form.

55

CRITICALLY DAMPED

We may obtain the solutions of interest quite quickly when $\zeta = 1$. Note that when $\zeta \rightarrow 1$, $\omega_d = \omega\sqrt{(1 - \zeta^2)} \rightarrow 0$. The force in the pile is then

$$F_p = K\left(\frac{\dot{\Delta}_0}{\omega_d} e^{-\zeta\omega t} \sin \omega_d t\right) \rightarrow \frac{K\dot{\Delta}_0}{\omega_d} e^{-\omega t}(\omega_d t)$$

because $\sin \omega_d t \rightarrow \omega_d t$. Hence

$$F_p = K\dot{\Delta}_0 t e^{-\omega t} \qquad \zeta = 1$$

and, similarly, the pile-head displacement is

$$\Delta_2(t) = \frac{2\dot{\Delta}_0}{\omega}[1 - e^{-\omega t}(1 + \omega t)] \qquad \zeta = 1$$

These solutions are not oscillatory: they hold for $t < 2L/V_d$. The peak pile force occurs when $t = 1/\omega$ and is

$$F_{p,max} = \frac{K\dot{\Delta}_0}{\omega} e^{-1}$$

Note that $F_{p,max}$ will not have had a chance to occur before the reflected wave returns to the pile head if $\omega < V_d/2L$.

OVERDAMPED

In this case $\zeta > 1$, and we shall see presently that this situation does occur in pile-driving problems. When the system is overdamped the character of the solution changes: the circular trigonometric functions, which are periodic, are replaced by exponential functions, which are not. The solution may be shown to be

$$F_p = K\left[\frac{\dot{\Delta}_0}{\omega\sqrt{(\zeta^2 - 1)}} e^{-\zeta\omega t}\left(\frac{\exp[\omega\sqrt{(\zeta^2 - 1)}t] - \exp[-\omega\sqrt{(\zeta^2 - 1)}t]}{2}\right)\right] \qquad \zeta > 1$$

and

$$\Delta_2(t) = \frac{2\zeta\dot{\Delta}_0}{\omega}\left\{1 - e^{-\zeta\omega t}\left[\left(\frac{\exp[\omega\sqrt{(\zeta^2 - 1)}t] + \exp[-\omega\sqrt{(\zeta^2 - 1)}t]}{2}\right)\right.\right.$$

$$\left.\left. + \frac{\zeta}{\sqrt{(\zeta^2 - 1)}}\left(\frac{\exp[\omega\sqrt{(\zeta^2 - 1)}t] - \exp[-\omega\sqrt{(\zeta^2 - 1)}t]}{2}\right)\right]\right\} \qquad \zeta > 1$$

In this form the similarity between underdamped and overdamped solutions is preserved to some degree. However, it is helpful to recall that the hyperbolic functions are defined by

$$\sinh \theta = (e^\theta - e^{-\theta})/2, \qquad \cosh \theta = (e^\theta + e^{-\theta})/2$$

so that the solutions can be written more compactly as

$$F_p = K\left(\frac{\dot{\Delta}_0}{\omega\sqrt{(\zeta^2 - 1)}}\, e^{-\zeta\omega t} \sinh\left[\omega\sqrt{(\zeta^2 - 1)}t\right]\right) \qquad \zeta > 1$$

and

$$\Delta_2(t) = \frac{2\zeta\dot{\Delta}_0}{\omega}\left[1 - e^{-\zeta\omega t}\left(\cosh\left[\omega\sqrt{(\zeta^2 - 1)}t\right] + \frac{\zeta}{\sqrt{(\zeta^2 - 1)}}\sinh\left[\omega\sqrt{(\zeta^2 - 1)}t\right]\right)\right]$$

$$\zeta > 1$$

In this form the analogy with the underdamped case is obvious. Because the solutions are not oscillatory, the time of validity of the solution has the sole requirement of $t < 2L/V_d$.

Before moving to a worked example we should make one final point. The damping parameter ζ represents one way to measure the stiffness of the capblock relative to the pile. For example, when the capblock stiffness is great, $\zeta \gg 1$, and in this situation the peak pile force is

$$F_{p,max} \rightarrow K\dot{\Delta}_0/2\omega\zeta$$

or

$$F_{p,max} \rightarrow \rho A V_d \dot{\Delta}_0$$

This is the classical result (due to one of the originators of stress wave studies – namely, the 19th-century physicist Hopkinson) for the force a wire offers in arresting a falling weight. This result obtains when the capblock is infinitely stiff. The theory for it will be presented shortly.

On the other hand, when $\zeta \ll 1$ we have a capblock that is flexible compared to the pile and the result is

$$F_{p,max} \rightarrow K\dot{\Delta}_0/\omega$$

In this situation the pile may be considered rigid and $F_{p,max}$ is simply the reaction to a mass falling and being arrested by a spring (the capblock) of stiffness K. The fundamentally different forms for the time history of axial force in the pile head are shown schematically in Figure 2.13.

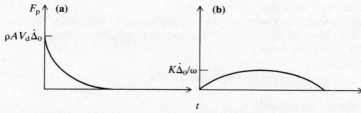

Figure 2.13 Limiting cases for dynamic loads in piles: (a) rigid capblock; (b) rigid pile.

57

FIRST WORKED EXAMPLE

The worked example has the following details: an offshore platform foundation pile of length 75 m is to be driven through mud to get to a sand layer 50 m below the mudline. The pile is a steel tube of outer diameter 1100 mm and wall thickness 35 mm, with $E = 2 \times 10^5$ MPa. It is driven by a steam hammer of ramweight 30 kN and energy 45 kN m (i.e. the ram falls 1.5 m). The capblock material is softwood, with a rated spring constant in compression of 10^3 kN mm^{-1}. The peak compressive pile force is to be calculated, within the limits of the theory outlined previously. In addition, the sequence of events arising when the capblock is omitted, and there is no cushion between ram and pile, is to be investigated. In this situation the ram stiffness is assumed to be infinite, which, of course, cannot be quite the case.

Data are:

$$A = \pi \times 1065 \times 35 \times 10^{-6} = 0.117 \text{ m}^2$$
$$\rho = 7.85 \text{ kN s}^2\text{m}^{-4}$$
$$E = 2 \times 10^8 \text{ kN m}^{-2}$$
$$V_d = \sqrt{(E/\rho)} = 5.05 \times 10^3 \text{ m s}^{-1}$$
$$\omega = \sqrt{(K/M)} = 5.72 \times 10^2 \text{ rad s}^{-1}$$
$$\dot{\Delta}_0 = \sqrt{(2g \times 1.5)} = 5.42 \text{ m s}^{-1}$$
$$\zeta = \sqrt{(KM)}/2\rho A V_d = 0.189$$
$$\omega_d = \omega\sqrt{(1 - \zeta^2)} = 5.62 \times 10^2 \text{ rad s}^{-1}$$

Thus, the system is underdamped because $\zeta < 1$. The time of validity of the solution is

$$t < \pi/\omega_d < 0.0056 \text{ s}$$

the other requirement being less stringent in this instance. Hence, as can be seen, the stresses are generated very quickly in the pile (the peak value occurs by about 0.003 s). In fact, they occur long before the pile has a chance to respond to the blow by sinking into the soil: this differentiation of timescales is quite apparent from watching piling operations.

Figures 2.14a and b show the time histories of response for pile force and pile-head displacement. The peak values are 7250 kN and 5.5 mm, respectively. Note that the peak pile force corresponds to an axial stress of 62 MPa, which is quite low for steel and may be attributed directly to the relatively soft capblock. Notice the similarity between Figures 2.13b and 2.14a. Omitting the capblock makes a huge difference as we now see. It is tantamount to assuming the capblock is infinitely stiff, that is, ζ is infinitely large.

When no capblock is present, our equations simplify considerably. We now make no distinction between Δ_1 and Δ_2. The equation now governs response of a mass plus dashpot. We can write

$$M\ddot{\Delta} = -\rho A V_d\dot{\Delta}$$

which integrates immediately to

$$M\dot{\Delta} = -\rho A V_d\Delta + M\dot{\Delta}_0$$

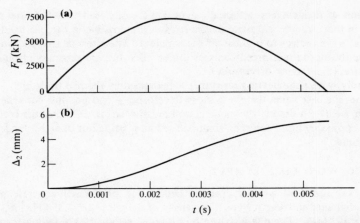

Figure 2.14 Time histories of response for an offshore pile drive by a steam hammer with a softwood capblock: (a) axial force in the pile head; (b) axial displacement in the pile head. (Note that solution is valid up to 0.0056 s.)

and so we obtain on a further integration

$$\Delta(t) = \frac{M\dot{\Delta}_0}{\rho A V_d} \{1 - \exp[-(\rho A V_d t)/M]\}$$

as the solution for the pile-head displacement. The pile force is

$$F_p = \rho A V_d \dot{\Delta} = \rho A V_d \dot{\Delta}_0 \exp[-(\rho A V_d t)/M]$$

the peak value of which occurs at the start and is

$$F_{p,max} = \rho A V_d \dot{\Delta}_0 = 25\ 000 \text{ kN}$$

leading to a peak axial stress of

$$\sigma_{max} = F_{p,max}/A = \rho V_d \dot{\Delta}_0$$

$$= 215 \text{ MPa}$$

This is, in essence, the analysis done originally by Hopkinson that was referred to earlier. Provided that the mass of ram is substantial, it is a curious fact that the stress generated (which in this case is high, but still below the yield strength of, say, 250 MPa) is independent of the mass: it depends only on the velocity of impact. The compliance of the ram itself will reduce this value of 215 MPa.

Notice that we have said nothing about the set of the pile due to each successive blow, nor have we discussed the energy balance upon which this obviously depends. These matters are not at all straightforward, and fairly wide differences of opinion exist.

As to the question of reflections we will bypass the problem, offering only qualitative comments. First, if the bearing resistance experienced by the pile tip is low, the initial compressive stress wave will be reflected as a tensile wave, and the

f incident and reflected waves could easily lead to substantial tensile
at vicinity. Likewise tensile stresses are possible in the pile head when
ave returns to bounce off the capblock. These tensile stresses can be of
cance in reinforced-concrete driven piles, but are generally insignificant
ssed-concrete driven piles.
, where the bearing resistance experienced by the pile tip is high − as is
some..... the case when the pile is close to refusing − it is possible for reflections
at the tip nearly to double the incident compressive stress. Pile damage from this
source is probably more common than realised and, being out of sight, tends to be
out of mind.

SECOND WORKED EXAMPLE

The second and final worked example concerns a prestressed concrete pile of
450 mm diameter and length 20 m. The density of the pile is $\rho = 2.35$ kN s^2 m^{-4} with
$E = 2.5 \times 10^4$ MPa. The pile is driven by a ram of weight 50 kN falling through a
height of 0.3 m. A softwood capblock with a rated stiffness in compression of
$K = 1.5 \times 10^3$ kN mm^{-1} is used.

Data are:

$$A = \pi \times 0.45 \times 0.45/4 = 0.159 \text{ m}^2$$
$$\rho = 2.35 \text{ kN s}^2 \text{ m}^{-4}$$
$$E = 2.5 \times 10^7 \text{ kN m}^{-2}$$
$$V_d = \sqrt{(E/\rho)} = 3.26 \times 10^3 \text{ m s}^{-1}$$
$$\omega = \sqrt{(K/M)} = 5.42 \times 10^2 \text{ rad s}^{-1}$$
$$\dot{\Delta}_0 = \sqrt{(2g \times 0.3)} = 2.42 \text{ m s}^{-1}$$
$$\zeta = \sqrt{(KM)}/2\rho A V_d = 1.14$$

Thus, the system is overdamped because $\zeta > 1$. The time of validity of the
solution reverts to

$$t < 2L/V_d < 0.0123 \text{ s}$$

which is the time taken for the initial stress wave to return to the pile head. The other
time constraint has no meaning in this instance because the solution is not
oscillatory.

The solutions are shown plotted in Figure 2.15a and b. Notice the significant
differences in form and magnitudes compared with Figure 2.14. Peak values are
2200 kN and 9 mm, respectively. The peak axial stress in the pile is 14 MPa, which
is reasonable for prestressed concrete. Notice, however, the similarity between
Figures 2.13a and 2.15a.

2.2.2 Suddenly applied load of constant magnitude but finite duration

An obvious extension of the preceding work is to limit the duration of the
applied load. It will be recalled that one of the objects in calculating the
period of the 25-storey building of Example 1.3 was to see if the duration
of the gust loading was sufficiently long for the peak response to develop.

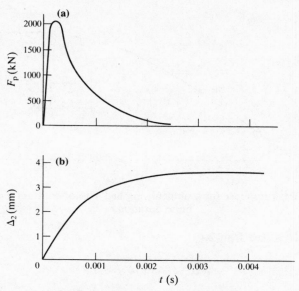

Figure 2.15 Time histories of response for a prestressed concrete pile driven by a steam hammer with a firm capblock: (a) axial force in the pile head; (b) axial displacement in the pile head.

Therefore, let us consider the case when the applied loading is in place only for a time t_*, and let us ignore damping. From the work of the previous section we can state categorically that if $t_* > T/2$, the peak response occurs during the application of the force. That is

$$\Delta_{\max} = 2F/K \qquad t_* \geqslant T/2 \qquad (2.17)$$

Thus, there is no point in any further consideration of step loading when its duration is greater than half the period of the system: peak response is given by Equation 2.17.

Because peak response occurs during the pulse when $t_* \geqslant T/2$, it follows that peak response occurs after the pulse when $t_* < T/2$. Thus, in determining peak response we have an initial-value problem to solve because there is no loading, only an initial displacement and an initial velocity communicated at time $t = t_*$.

For free vibrations with initial conditions of

$$\Delta_0 = \frac{F}{K} (1 - \cos \omega t_*)$$

and

$$\dot{\Delta}_0 = \frac{\omega F}{K} \sin \omega t_*$$

61

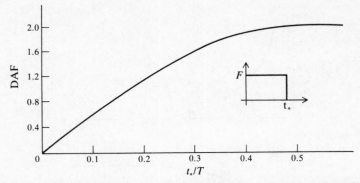

Figure 2.16 Peak response for a suddenly applied load of constant magnitude but finite duration.

the amplitude is (see Eqn 2.4)

$$\Delta_{max} = \sqrt{[\Delta_0^2 + (\dot{\Delta}_0/\omega)^2]}$$

or

$$\Delta_{max} = \frac{\sqrt{2}F}{K} \sqrt{(1 - \cos \omega t_*)} = \frac{2F}{K} \sin(\pi t_*/T) \qquad t_* \leqslant T/2 \qquad (2.18)$$

Note that we recover our earlier result of $\Delta_{max} = 2F/K$ when $t_* = T/2$. Equations 2.17 and 2.18 are shown plotted in Figure 2.16. The ordinate, DAF, is the dynamic amplification factor, which we define as

$$DAF = \Delta_{max}/(F/K)$$

if we think in terms of displacement or, also

$$DAF = (K\Delta_{max})/F$$

if we think in terms of force.

2.2.3 Other simple loading cases

The dynamic amplification factors for another three common simple loading cases are shown in Figures 2.17–19. These are based on work presented by Biggs(1964).

A word or two of comment is in order. Figure 2.17 shows response when the applied load rises to a constant value. The abscissa is appropriately labelled t_*/T, where t_* is this rise time. When t_* is small, say $t_* < 0.25T$, the reduction in the DAF below the value 2 is insignificant and this thus provides a semiquantitative definition of the term 'suddenly applied' – that

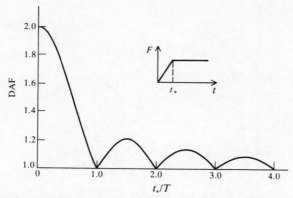

Figure 2.17   Peak response for a constant load with finite rise time (after Biggs (1964)).

is, if the time required for the applied load to reach its peak intensity is less than about 25% of the system period we may, for all practical purposes, assume that the loading comes on instantaneously.

It is a peculiarity of the system that when t_* is an integer multiple of the period of the structure the peak response is simply the static response. Furthermore, it is apparent that the static solution is recovered, inevitably, when t_* is large. This is hardly surprising since a substantial rise time is consistent with the notion of static loading (see Problem 2.8).

Turning to Figure 2.18, in which response to a triangular load is depicted, we note that it should be coupled with Figure 2.16. In both Figures 2.16 and 2.18 the load is truly suddenly applied: response to the triangular loading is always less severe, however, for the very good reason that less load is applied. Nevertheless, peak values of the DAF equal to 2 in one case, and approaching it in the other, are recorded when the duration of the load becomes lengthy.

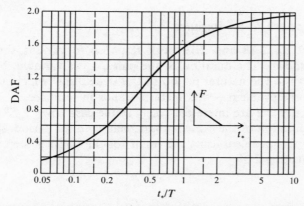

Figure 2.18   Peak response to a suddenly applied triangular load block (after Biggs (1964)).

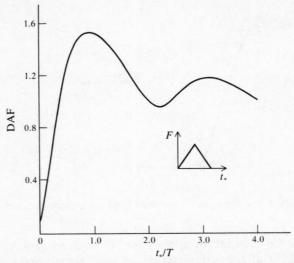

Figure 2.19 Peak response to a symmetric 'saw tooth' load block (after Biggs (1964)).

In Figure 2.19 response to a single 'saw tooth' pulse is depicted and the two items of note are that the static solution is once again recovered when the time of application is large, and that the maximum value of the DAF is about 1.5.

Between them Figures 2.16 through 2.19 contain a lot of information useful to the practitioner who needs to go a step beyond a 'back of the envelope' calculation. We could have presented all the theory on which the curves are based but, quite frankly, there is not much point: the numerical techniques outlined in Chapter 4 provide an inkling of the most sensible route to follow when a lot of data have to be handled.

2.2.4 *Response to an impulse*

In Figures 2.16, 2.18 and 2.19 peak response to various simple pulse loadings is shown. The duration of the pulse, t_*, is variable, but it is its length relative to the natural period of the structure – which is the other timescale in the problem – that dictates trends in response.

In this section we are interested solely in the case when t_*/T is small – that is, the duration of the pulse is short relative to the period. It is easiest to approach this problem using the result for the uniform step loading as given by Equation 2.18.

We have

$$K\Delta_{max}/F = 2 \sin(\pi t_*/T)$$

which holds for $t_*/T \leqslant 1/2$. Now when $t_*/T \ll 1$, $\sin(\pi t_*/T) \approx \pi t_* T$. Thus as t_*/T becomes small

$$K\Delta_{max}/F \to 2\pi t_*/T$$

or

$$\Delta_{max} \to \omega(Ft_*)/K \qquad (2.19)$$

Clearly, peak response is directly related to the duration of the pulse. The formation of the product Ft_* is deliberate since this is the impulse, I, for the present loading.

More generally, we may define the impulse as

$$I = \int_0^{t_*} F(t)\, \mathrm{d}t \qquad \text{for } t_*/T \ll 1$$

For most structures a hammer blow is a good example of an impulse. The peak applied force may be high, but its duration is decidedly short-lived. The example treated earlier in this chapter of airplane impact on a tall building is certainly of this nature, so we have already skirted around the issue of response to an impulse. In a formal sense, therefore, the problem of response to an impulse is one of free vibrations under imposed initial conditions. An impulse cannot generate an initial displacement. However, there is an initial velocity generated by an impulse. Newton's second law is

$$\frac{d}{dt}(MV) = F$$

which, in present terminology, after integration becomes

$$M\dot{\Delta}_0 = \int_0^{t_*} F(t)\, \mathrm{d}t = I \qquad (2.20)$$

Thus, there is a requirement for initial conditions of the form $\Delta_0 = 0$, $\dot{\Delta}_0 = I/M$, so that the undamped response to an impulse is

$$\Delta(t) = \frac{\dot{\Delta}_0}{\omega} \sin \omega t = \frac{I}{M\omega} \sin \omega t \qquad (2.21)$$

and the peak response is

$$\Delta_{max} = \frac{I}{M\omega} = \frac{I\omega}{K} \qquad (2.22)$$

from which we recover the earlier, less general, result given by Equation 2.19.

Incidentally, in the two cases of triangular impulses (refer to Figs 2.18 & 2.19), the area of each triangle is $\frac{1}{2}Ft_*$ and the peak response is

$$\Delta_{max} = \tfrac{1}{2}(Ft_*)\omega/K$$

This is borne out by an inspection of the relevant regions of Figures 2.18 and 2.19. However, the crux of the matter – and it is of some significance in a practical sense – is that for loadings of short duration the actual variation in applied load need not be known or, alternatively, need not be accounted for explicitly. The integrated quantity – the impulse – based as it is on the concept of momentum, provides the most expeditious go-between between forcing and response (see, for example, Examples 4.2 and 4.3).

Example 2.4   The equivalent force of impact

The method used in Example 2.1 to calculate the equivalent force generated by the impact of the aircraft is not particularly accurate. It is better to use Newton's second law directly. The time of impact is roughly the length of the airliner divided by its velocity. If the length is of the order of 70 m and the velocity is 300 km h^{-1} this time, Δt, say, is $\Delta t \approx 5/6$ s. This is a small fraction of the period of the structure, which was estimated to be around 8–9 s. Hence the equivalent force is decidedly impulsive and of magnitude

$$F = \frac{\Delta (MV)}{\Delta t} \approx \frac{2000}{9.81} \times \frac{300 \times 10^3}{3600} \times \frac{6}{5} \approx 20\ 000 \text{ kN}$$

Thus the deceleration is of the order of $10g$, which adds a final chilling note to a rather macabre example.

2.3   Elasto-plastic response to a suddenly applied load of constant magnitude

If a structure is to be designed to resist a suddenly applied load, F, which is held constant for an appreciable period, and that design is to be carried out under the assumption of working stresses, then the load to be resisted elastically is $2F$. In many instances this will require a substantial structure, but deflections will remain within the elastic region.

On the other hand, if the structure is permitted to yield and if the resulting deflections are not unacceptably large, a less substantial structure is required, which has an obvious attendant economic advantage.

This concept, which is to dynamic analysis what plastic theory is to static analysis, has far-reaching consequences. It lies at the heart of much of modern practice in earthquake engineering and the design of structures to resist impact and blast loadings. It is a somewhat unpalatable fact, but a fact nonetheless, that it is often impossible to design a structure to resist, solelv by elastic means, unexpected and not easily quantifiable loads. The loads are sometimes too great, and the only available route is to provide less resistance, permit the structure to yield in a controlled manner and so absorb the energy of the impact by plastic deformation.

It should be clear that we have to date considered only systems in which the resistance is linear and elastic and as such represented by a spring of constant stiffness. However, the stiffness is constant only as long as the structure remains in the elastic region. In a ductile structure (for example, well detailed mild steel or well detailed reinforced concrete, in both of which permitted strains may be several times the yield strains), a reasonable curve of resistance versus deflection, for increasing deflection, is the elasto-plastic curve shown in Figure 2.20. The kink in the curve isolates the point at which the elastic capacity of the structure is consumed, after which increased resistance cannot be supplied.

If damping is ignored − it is often of little consequence compared to the dissipation due to plastic action − we may write the energy equation for a suddenly applied constant load as

$$\tfrac{1}{2}M\dot{\Delta}^2 + \int_0^\Delta R(\Delta)\,d\Delta = F\Delta \tag{2.23}$$

where $R(\Delta)$ is the resistance function. For a linearly elastic system $R \equiv K\,\Delta$ and the integral is then simply $\tfrac{1}{2}K\,\Delta^2$, as expected.

Now, up to the time of maximum displacement response, $\Delta(t)$ increases steadily and $\dot{\Delta}(t)$ is positive. For a structure that enters the plastic region, the elasto-plastic behaviour illustrated in Figure 2.20 permits the integral in Equation 2.23 to be evaluated explicitly, namely

$$\int_0^\Delta R(\Delta)\,d\Delta = \tfrac{1}{2}R\Delta_{el} + R(\Delta - \Delta_{el})$$

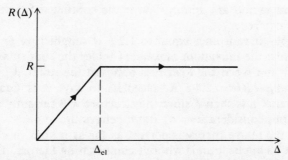

Figure 2.20 Elasto-plastic behaviour for increasing deflection.

where R is the resistance supplied and Δ_{el} is the displacement at which the elastic limit is reached, that is, $\Delta_{el} = R/K$. In a physical sense the integral is obviously the area under the resistance–displacement curve.

Consequently, the energy equation becomes

$$\tfrac{1}{2} M \dot{\Delta}^2 + \tfrac{1}{2} R \Delta_{el} + R(\Delta - \Delta_{el}) = F\Delta \qquad (2.24)$$

We are interested only in the maximum value of Δ generated by F. This occurs when $\dot{\Delta}$ is zero, that is, the kinetic energy is momentarily zero, and so

$$\tfrac{1}{2} R\Delta_{el} + R(\Delta_{max} - \Delta_{el}) = F\Delta_{max}$$

Rearranging this gives

$$R/F = 2\mu/(2\mu - 1) \qquad (2.25)$$

where $\mu = \Delta_{max}/\Delta_{el}$ is the ratio of full plastic displacement to elastic displacement. This is usually called the ductility factor. Equation 2.25 holds only for $\mu \geqslant 1$.

When $\mu = 1$

$$R = 2F$$

and we have to supply the maximum amount of resistance. However, when μ is large

$$R \to F$$

and we need supply only half as much.

In practice a value of $\mu = 10$ would be considered large, implying appreciable damage and the possibility of expensive repair bills. Therefore, large permitted values of μ imply that the structure or structural element can be sacrificed, but they cannot be so high that collapse would occur. On the other hand more moderate values of $\mu = 4$ to 6 will, in many cases, lead to minimal damage and are appropriate if the continued functioning of the structure is required.

For example, if resistance equal to $1.2\,F$ is supplied we find that $\mu = 3$. In other words, the maximum movement under the sudden application of F is $3\,\Delta_{el}$. If and when the load F is removed the residual permanent set would be $3\,\Delta_{el} - R/K = 2.4\,F/K$. Needless to say, it is unacceptable to supply resistance less than F since the structure will certainly collapse – as is evident from considerations of static behaviour.

Although the above theory is correct as far as it goes, it should not be inferred that its application is without complicating factors. There are, for example, serviceability requirements for other loading conditions which may alter, profoundly, the constraints that can be imposed. In any event, when the loading is sudden, and of appreciable duration, the capacity required of the structure is still appreciable, although significant savings may be realised *vis-à-vis* a conventional working stress approach.

At the other extreme we could consider a suddenly applied load of appreciable magnitude but very short time of occurrence. This is of interest on two counts, at least: first, such loadings are quite common, e.g. blast and impact; secondly, any reasonable variation with time of the applied force is acceptable because we shall invoke the concept of the impulse, the integrated quantity that smoothes out such variation. In other words, with respect to this latter point, the theory at this end of the spectrum is somewhat more general than that implied by Equation 2.25, which is for the case of a uniform load only. Note that we do not invoke Equation 2.24 because the forcing is over very quickly. In fact, over this short time

$$\int F \, d\Delta = \int M \ddot{\Delta} \, \frac{d\Delta}{dt} \, dt$$

$$= \tfrac{1}{2} M \int d(\dot{\Delta})^2$$

$$= \tfrac{1}{2} M \dot{\Delta}_0^2$$

So, under an impulse I (which gives rise to an initial velocity $\dot{\Delta}_0 = I/M$) the kinetic energy transmitted to the system (almost instantaneously) is

$$\int F \, d\Delta = \tfrac{1}{2} M \dot{\Delta}_0^2 = \tfrac{1}{2} I^2/M$$

At the time of maximum displacement, all this kinetic energy has been transferred to elastic strain energy and plastic dissipation. That is

$$\tfrac{1}{2} I^2/M = \tfrac{1}{2} R \Delta_{el} + R(\Delta_{max} - \Delta_{el})$$

or

$$R/I\omega = 1/\sqrt{(2\mu - 1)} \qquad (2.26)$$

Again, when $\mu = 1$, we have the elastic theory result. But when μ is large we see that the required resistance can be quite small in comparison to the 'input' $I\omega$. Example 2.5 picks up on this.

The final matter to consider, before we look at four examples, is to enquire as to what simplification can be brought to bear on the case where the structure, although adequately modelled as a single-degree-of-freedom system, is multiply redundant. In such cases the resistance–displacement curve must take account of several plastic hinges, with the associated reduction in stiffness that accompanies the formation of each hinge.

69

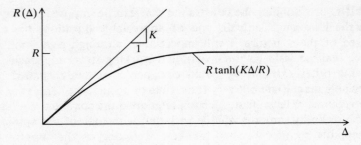

Figure 2.21 Idealised elasto-plastic behaviour for multiply redundant systems.

Obviously there will be an initial elastic portion and there will be a final horizontal tail corresponding to a full plastic mechanism and, therefore, an upper limit on resistance. In between there will be a curve which it is not unreasonable to assume is of the form (see Fig. 2.21).

$$R(\Delta) = R \tanh(K \Delta / R) \qquad (2.27)$$

The hyperbolic tangent function has the desired characteristics of $R(\Delta) \to K \Delta$ when Δ is small (i.e. linearly elastic behaviour), and $R(\Delta) \to R$ when Δ is more appreciable (i.e. the horizontal tail is traversed). The integral we require is

$$\int_0^\Delta R(\Delta) \, d\Delta = R \int_0^\Delta \tanh(K \Delta / R) \, d\Delta$$

$$= (R^2/K) \ln[\cosh(K \Delta / R)]$$

For the case of a constant force F, suddenly applied, we have

$$F \Delta_{max} = (R^2/K) \ln[\cosh(K \Delta_{max}/R)] \qquad (2.28)$$

from which R must be found implicitly once Δ_{max} is prescribed (and vice versa). For the impulsive loading, we have

$$\tfrac{1}{2} I^2 \omega^2 = R^2 \ln[\cosh(K \Delta_{max}/R)] \qquad (2.29)$$

and, again, an implicit solution is required. In both cases, notice that it is difficult to define a ductility factor, because the displacement at the elastic limit cannot be clearly defined. In practice one is usually able to determine the load at which the first hinge forms in a multiply redundant system: in many instances this would qualify as the elastic limit of the structure, but the choice rests with the practitioner.

Example 2.5 Rock fall on a protective cage

A conveyor belt passes over another belt in an ore handling plant. In order to prevent damage to the lower conveyor from a rock fall from the upper belt, a simple protective frame is recommended. The principal structural element is a simply supported steel beam, of length 5 m, which is laterally braced so that its full moment capacity can be mobilised. The details are shown in Figure 2.22.

The rock, of weight 10 kN, can fall through 5 m, and to maintain adequate clearance the peak midspan deflection is limited to $\Delta_{max} = 0.125$ m.

In order quickly to obtain the required size we shall ignore beam self-weight and the contribution of elastic strain energy – these, as can readily be shown subsequently, are of minor significance in this problem. Since the beam deflects by Δ_{max} the potential energy lost by the rock is $W(h + \Delta_{max})$ and this must equal the energy dissipated by the midspan plastic hinge in the beam. That is

$$R\Delta_{max} = W(h + \Delta_{max})$$

From beam theory

$$R = 4M_p/L$$

where M_p is the plastic moment of resistance of the beam and L is the span. Hence

$$M_p = \frac{WL}{4}\left(1 + \frac{h}{\Delta_{max}}\right)$$

$$= \frac{10 \times 5}{4}\left(1 + \frac{5}{0.125}\right)$$

$$= 5.125 \times 10^2 \text{ kN m}$$

For normal structural-grade steel the required beam is a 530 UB 82. The deflection at which the elastic limit is reached for this beam is about 10 mm, so the required ductility factor is $\mu = 125/10 \approx 12$: this is achievable.

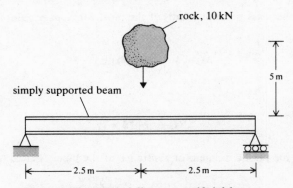

rock, 10 kN

5 m

simply supported beam

├── 2.5 m ──┤├── 2.5 m ──┤

Figure 2.22 Rock fall on a sacrificial beam.

71

It must be mentioned in passing that it is not possible to find a large enough readily available rolled section to do the job if elastic behaviour is stipulated. A plastic approach is the only sensible route to follow.

Example 2.6 Pipe fracture and subsequent pipe whip

To set the scene a quotation from Peek is in order.

> In the event of rupture of a high pressure pipeline the jet of fluid (steam or water) emanating from the rupture causes a large thrust. Since the pipe supports are generally not designed to withstand this load, additional pipe whip restraints must be provided to prevent the ruptured pipe from accelerating and striking other pipeline or essential instruments. To enable unrestrained thermal and seismic motions during operating conditions, a gap is provided between these restraints and the pipe [see Fig. 2.23]. After rupture, the pipe accelerates through this gap, thus gaining a large amount of kinetic energy which must be absorbed by the restraint.

Under the assumption of rigid–plastic response (that is, elastic effects are ignored) sudden fracture of the pipe near a bend, say, mobilises a thrust F due to the fluid pressure. This thrust is resisted by inertia forces over a length l of pipe to the point where a plastic hinge forms. One assumes that the inertia forces are linearly distributed because the pipe is assumed to rotate as a rigid body about the hinge position. The plastic hinge forms where moment is a maximum, which, in the present case, implies that it forms where the shear force is first zero (see Fig. 2.24).

The two relevant equilibrium equations are shear and moment equilibrium, and they may be used to find the two unknown quantities, which are the length of pipe to the plastic hinge, l, and the acceleration of the tip, $\ddot{\Delta}$. Shear force equilibrium yields

$$\ddot{\Delta} \int_0^l m(x/l)\,\mathrm{d}x = F$$

or

$$\tfrac{1}{2} m l \, \ddot{\Delta} = F$$

where m is the mass per unit length of the pipe. Moment equilibrium yields

$$M_\mathrm{p} + \ddot{\Delta} \int_0^l m(x/l)x\,\mathrm{d}x = Fl$$

or

$$M_\mathrm{p} + \tfrac{1}{3} m l^2 \ddot{\Delta} = Fl$$

where M_p is the plastic moment of resistance of the pipe. Hence,

$$l = 3M_\mathrm{p}/F$$

and so

$$\ddot{\Delta} = \frac{2F^2}{3mM_p}$$

Thus the displacement Δ at time t after rupture (assuming the force F remains constant) is

$$\Delta = \frac{F^2 t^2}{3mM_p}$$

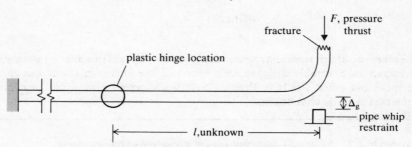

Figure 2.23 An idealised pipe at fracture.

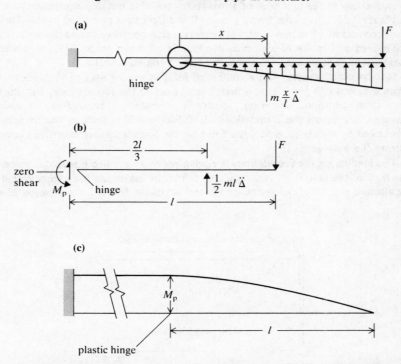

Figure 2.24 Conditions after fracture and prior to impact on restraint element: (a) resisting inertia forces based on rigid-body rotation about hinge; (b) free-body diagram; (c) bending moment diagram.

If a displacement Δ_g is traversed before the pipe bears on the restraint, the work done by the thrust is

$$F\Delta_g$$

while the energy absorbed at the plastic hinge is

$$M_p(\Delta_g/l)$$

The ratio is

$$\frac{M_p(\Delta_g/l)}{F\,\Delta_g} = \frac{M_p}{Fl} = \frac{1}{3}$$

The other two-thirds represents kinetic energy of the pipe and this must be dissipated by impact on a suitably designed whip restraint. The interested reader can chase up these and other details in Peek's (1982) work, where a thorough physically motivated study is undertaken.

Example 2.7 Effect of blasting on an elevated machine floor

A typical bay of the first floor of a two-storey building housing equipment is shown in Figure 2.25. The upper frame (omitted) is a light bent providing cover. The first floor consists of a 150 mm reinforced concrete slab on stringers and the main frames are spaced at 4 m. Dead load plus live load is estimated to be 30 kN m^{-1} and the majority of the live load is always in place, being equipment.

It is proposed to carry out a controlled blast in nearby rock. The pressure wave from the blast will give rise to a high lateral load on the foundations, but this is of very short duration. The wave velocity is estimated to be 1000 m s^{-1}, giving a passage time across the foundations of $10/1000 = 0.01$ s. Because the foundations are linked by a slab on grade, the force on the foundations is essentially constant during the passage.

The loading on the foundations is equivalent to an applied horizontal force F at the level of the first floor. It is required to find the maximum blast force F that can be allowed given that the lateral permanent set of the frame cannot exceed 25 mm.

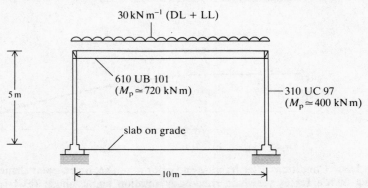

Figure 2.25 Definition diagram for portal frame.

(a)

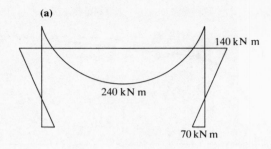

140 kN m

240 kN m

70 kN m

(b)

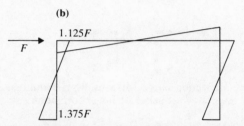

F

1.125F

1.375F

Figure 2.26 Various bending moment diagrams: (a) under dead load plus live load; (b) under load F.

The bending moment diagram under vertical dead load plus live load is shown in Figure 2.26a. Under small lateral loads applied to the knee of the portal frame, the bending moment diagram for this loading is as shown in Figure 2.26b.

The elastic lateral stiffness of the frame is calculated to be

$$K = F/\Delta = 8.6 \text{ kN mm}^{-1}$$

The collapse mechanism is, in this case, a sideways mechanism with hinges forming at the top and bottom of each column. The resistance is

$$R = 4M_p/5$$

But, for the 310 UC 97 columns $M_p \approx 400$ kN m and so

$$R = 320 \text{ kN}$$

Further analysis shows that, owing to the presence of the constant vertical loading, the sequence of plastic hinge formation is as shown in Figure 2.27.

A permissible permanent set of 25 mm means that the peak lateral displacement is limited to

$$25 + R/K = 25 + 320/8.6$$

$$= 62 \text{ mm}$$

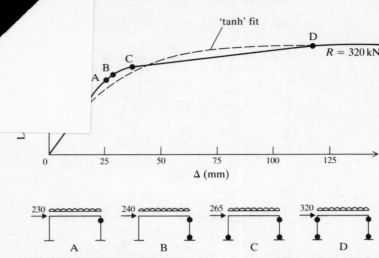

Figure 2.27 Resistance–deflection curve for a multiply redundant structure and the associated sequence of hinge formation.

which, by reference to Figure 2.27, means that three of the four hinges have formed: that is, a full mechanism has not formed. The broken curve in Figure 2.27 is the 'tanh' fit based on matching the initial slopes in each case. The agreement is good, with the area under each curve being more or less identical.

The natural period of the structure is

$$T = 2\pi \sqrt{\left(\frac{M}{K}\right)} = 2\pi \sqrt{\left(\frac{30 \times 10}{9.81 \times 8.6 \times 1000}\right)} = 0.37 \text{ s}$$

The natural circular frequency is $\omega = 16.8 \text{ rad s}^{-1}$. Our impulse formula for controlled plastic action in a multiply redundant structure is

$$\omega I = \omega F t_* = \sqrt{2R} \sqrt{\{\ln[\cosh(K\,\Delta/R)]\}}$$

Now $t_* = 0.01$ s, $\omega = 16.8 \text{ rad s}^{-1}$, $R = 320$ kN, $K = 8.6 \text{ kN mm}^{-1}$ and $\Delta = 62$ mm. Hence, the greatest blast force that can be permitted is

$$F = \frac{\sqrt{2} \times 320}{0.01 \times 16.8} \times 1.01 = 2700 \text{ kN}$$

or, in terms of peak acceleration applied,

$$a = \frac{F}{M} = \frac{2700}{300} g = 9g$$

The blast loading is most unlikely to be as high as this. In one case known to the author (where measurements were taken) involving a substantial charge set off about 20 m from an existing building, a peak acceleration pulse of $4g$ was recorded.

In the case of a blast acceleration of $4g$

$$\omega I = 16.8 \times 4 \times 9.81 \times \frac{300}{9.81} \times 0.01$$

$$= 200 \text{ kN}$$

However, $R = 320$ kN and we see from the previous figure that the first hinge forms at a lateral load of 230 kN. Hence the $4g$ pulse is not sufficient to push the frame into the inelastic region: the peak lateral displacement is $27 \times 200/230 = 24$ mm, which occurs at a time $T/4 = 0.09$ s after the blast hits and long after it has passed by.

Example 2.8 *Elasto-plastic action in a pile*

The last example in this chapter examines the situation with the steel pile considered earlier in Example 2.3 when the stroke and the ram are increased substantially. The ram is of weight 300 kN, the stroke, or fall, is now 3 m, and no capblock is used. The yield strength of the steel is taken to be 250 MPa.

Because a capblock is not used, the peak stress in the pile head occurs at the moment of impact. Under the assumption of elastic behaviour this stress is

$$\sigma = \rho V_d \dot{\Delta}_0$$

Since this is greater than the yield stress of the material, the initial phase of the response is purely plastic at a limiting stress of 250 MPa.

The equation of motion is

$$M\ddot{\Delta} = -F_p = -250 \times 10^3 \times 0.117$$

or

$$\frac{300}{9.81} \ddot{\Delta} = -2.93 \times 10^4 \text{ kN}$$

yielding

$$\ddot{\Delta} = -956 \text{ m s}^{-2}$$

Integrating once gives

$$\dot{\Delta} = -956\,t + \text{constant}$$

and the constant is the initial velocity of impact, namely 7.67 m s^{-1}, so

$$\dot{\Delta} = -956\,t + 7.67$$

Hence, the plastic displacement of the pile head is

$$\Delta = -478\,t^2 + 7.67\,t$$

77

This continues until $\dot{\Delta}$ is such that the axial stress in the elastic equation is the yield stress. That is, until

$$\dot{\Delta} = \sigma_y / \rho V_d$$

$$= \frac{250 \times 10^3}{7.85 \times 5.05 \times 10^3} = 6.3 \ \text{m s}^{-1}$$

which occurs at a time after impact of

$$t_* = \frac{7.67 - 6.3}{956} = 0.0014 \ \text{s}$$

The zone of plastification extends a distance

$$\Delta_* = 7.67 \times 0.0014 - 478 \times (0.0014)^2$$

$$= 0.010 \ \text{m}$$

into the pile head. After this the elastic wave takes over and propagates away at a velocity 'equal' to the speed of sound in the pile (for this region of interest look back at Example 2.3).

The sequence of events in terms of force in the pile and axial displacement is shown by the curves in Figure 2.28. Clearly, plastification leads to a significant jump in axial deflection.

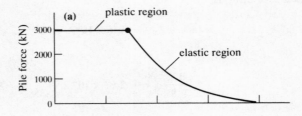

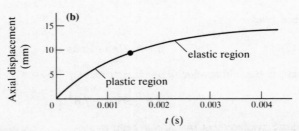

Figure 2.28 Plastic–elastic time histories for pile tip: (a) pile force; (b) axial displacement.

Problems

2.1 A tending vessel of weight 10 000 kN impacts a bracing member of a steel jacket with an initial velocity of 0.5 m s^{-1}. The member is a tube 15 m long, 40 cm in diameter, 15 mm in wall thickness with a yield stress of 250 MPa. The impact occurs at the waterline, which in this case corresponds approximately to the midspan of the member.

Assuming a plastic collapse mechanism forms instantaneously (i.e. rigid–plastic behaviour) with plastic hinges at each end joint and at midspan, find the constant retarding force offered against the vessel. Further, find by how much the member is kinked before the vessel comes to rest.

The above is best solved using energy principles, but use Newton's second law to establish that the time taken to bring the vessel to rest is approximately 1.6 s.

2.2 After sudden fracture near an elbow the thrust accelerating a portion of pipe perpendicular to its axis is 80 kN. The plastic moment of resistance of the pipe is 40 kN m and its weight per unit length is 0.3 kN m^{-1} (look back at Example 2.6).

Locate the plastic hinge and find the time it takes before the gap of 50 mm between pipe and whip restraint is closed. If, after hitting the whip restraint, the pipe moves another 40 mm before coming to rest, estimate the restraint strength required (based on perfectly plastic action).

It may be assumed that the thrust, which is due to high-pressure steam, remains constant throughout the impact process and that the pipe is of substantial length between bends.

2.3 The building described in Example 2.7 is subjected to a sequence of blasting in nearby rock. The pattern of loading is as shown in Figure 2.29.

Assuming the structure remains in the elastic region, use impulse theory and the principle of superposition to estimate the peak lateral displacement of the frame.

Armed with this knowledge, and by referring back to Figure 2.26, decide whether the assumption of linear elastic response can be justified in this instance.

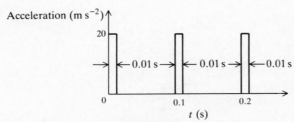

Figure 2.29

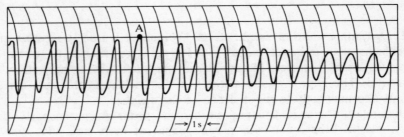

Figure 2.30

2.4 The trace shown in Figure 2.30 is of free vibrational decay for an eight-storey building. Point A on the diagram locates the time at which the eccentric mass exciter (located on the top floor of the building) was switched off and the previous steady-state response begins to decline in amplitude. Estimate the fraction of critical damping indicated.

2.5 This problem concerns idealised motion of an elevator (Fig. 2.31). A spring of stiffness K supports a weight of mass M. If the upper end begins to move upwards with a steady velocity of V, show that the distance Δ that the mass has risen in time t is governed by the equation

$$M\ddot{\Delta} + K\Delta = KVt$$

Select appropriate initial conditions and hence show that the solution is

$$\Delta(t) = Vt - (V/\omega)\sin\omega t$$

where $\omega = \sqrt{(K/M)}$. Plot the result and find the peak spring force due to this imposed motion. The similarity of this problem to that in Example 1.1(b) should be noted.

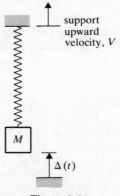

support
upward
velocity, V

M

$\Delta(t)$

Figure 2.31

2.6 A steel universal beam pile, 310 UBP, of area 0.01 m^2, is of length 45 m. It is driven by a ram of weight 35 kN falling through a height of 1.1 m. Using the theory presented in Example 2.3, find the peak axial compressive stress generated in the head of the pile when a capblock of stiffness $13\,000 \text{ kN mm}^{-1}$ is inserted.

2.7 Returning to Example 1.1(b), where the hoisting operation in an offshore crane was considered, we note that an alternative route to the solution for the peak axial force in the hoist rope should be possible using the principle of superposition and analytical solutions for free vibrations under prescribed initial conditions and response to a suddenly applied load. If we start the problem at the instant the payload lifts clear of the vessel then the initial conditions are $\Delta_0 = Mg/K$ and $\dot{\Delta}_0 = V_{\mathrm h}$ and, analytically speaking, the payload is at that instant suddenly applied − gravity is switched on. Hence show that the required solution is

$$\Delta(t) = (Mg/K) + (V_{\mathrm h}/\omega) \sin \omega t$$

and so establish the DNV formula. Seeing the result in this form suggests another interpretation which is perhaps the simplest of all: namely, that of an initial velocity imparted to a system which already has the static deflection accounted for.

2.8 Show that the solution for linearly increasing forcing of the form Ft/t_* is

$$\Delta = \frac{F}{K}\left(\frac{t}{t_*} - \frac{1}{\omega t_*} \sin \omega t\right), \qquad 0 \leqslant t \leqslant t_*$$

Show that, if t_* is a multiple of the period T, then $\dot{\Delta} = 0$ when $t = t_*$ and at that time

$$\Delta_{\max} = F/K$$

This helps explain Figure 2.17.

2.9(a) A weight of mass M falls through a height h and impacts a spring-supported receiving structure of mass m. If the impact is plastic (i.e. the two masses coalesce on impact and continue downward with a common velocity), show that the peak dynamic displacement, δ, is found from the solution of the equation

$$\tfrac{1}{2} K \delta^2 - \mathrm{Mg}\, \delta - \left(\frac{M}{M+m}\right) \mathrm{Mgh} = 0$$

where K is the spring stiffness of the supporting structure. The peak force transmitted to the foundation (including the weight of receiving structure) is $mg + K \delta$.

Note that if the receiving structure is massive a considerable proportion of the energy of the falling mass is lost on impact. This type of problem, has wide application in practice (see, for example, Problem 3.2).

2.9(b) A conveyor feeds logs into a chipping mill. The structure supporting the conveyor is a four leg portal frame and its weight is about 30 kN. The logs weigh 30 kN on average and fall through a height of 0.3 m from the feed tray to the conveyor. In order to reduce the peak foundation loads and wear and fear of the portal frame it is proposed to insert spiral springs in each leg at the foundation level. If the peak axial load in each leg due to the falling log is not to exceed 60 kN, find a suitable spring and show that its peak displacement is 0.06 m.

3

Resonance
and related matters

3.1 Steady-state response to sinusoidal forcing

We begin an important section, for the nature of the response to sinusoidal forcing is quite different from anything so far considered.

In all previous cases, whether of free vibrations or of response to sudden forces of simple form, the vibration occurs at the natural frequency of the system. The vibration also dies away due to damping and we are left with zero displacement in all cases except those in which a constant force remains, in which case the final displacement is none other than the static displacement due to that force.

With forcing that persists indefinitely and is of the form

$$F(t) = F \sin \Omega t$$

where Ω is the natural circular frequency of excitation, we may isolate two regions of interest. The first, which we shall treat later, concerns the situation when vibration gets under way in response to $F(t)$ starting, say, at time $t = 0$. The application of this forcing function in this way is sudden and the response of the spring–mass system reflects this in that part of the response is at the natural frequency of the system. However, damping attenuates this portion and, with time, it dies away altogether. This can be seen in Figure 3.1, where the forcing frequency Ω is significantly less than the natural frequency ω.

The other part of the solution, which is in effect also from time $t = 0$, takes over and is expressed by the general relationship

$$\Delta(t) = \Delta_{max} \sin(\Omega t - \phi)$$

where Δ_{max} and ϕ are constants. That is, after a suitable initial period, the transients (which arose because we started the forcing at a given time) die

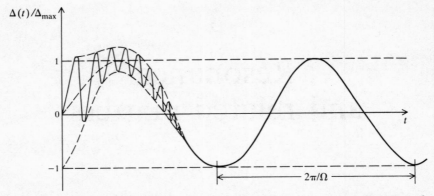

Figure 3.1 Attenuation of transient response leading to the development of a steady state.

away, and we are left with a solution that has precisely the same temporal character as the forcing function, insofar as the frequency of the response is identical to the frequency of the forcing (see the right-hand end of Fig. 3.1). The only differences are those of degree: the response is not, in general, in phase with the forcing (i.e. peaks in response do not correspond, timewise, to peaks in forcing); neither is the magnitude of the response related in a simple way to the magnitude of the forcing (response magnitude can be significantly greater or significantly smaller, for example).

This form of response is called the steady state and, with reference to the previous equation, it may be noted that Δ_{max} is the amplitude of the response and ϕ is the phase angle — the amount by which the response lags the input.

We are now in a position to proceed to a solution for steady-state response to sinusoidal forcing. The equation of motion is

$$\ddot{\Delta} + 2\zeta\omega\dot{\Delta} + \omega^2\Delta = (F/M)\sin \Omega t \qquad (3.1)$$

and the solution is given by

$$\Delta(t) = \Delta_{max} \sin(\Omega t - \phi) \qquad (3.2)$$

We find Δ_{max} and ϕ by substitution of the stipulated solution in the differential equation. First, however, we write

$$\sin \Omega t = \sin[(\Omega t - \phi) + \phi]$$

$$= \sin(\Omega t - \phi)\cos \phi + \cos(\Omega t - \phi)\sin \phi$$

84

On differentiation, substitution and rearrangement we have

$$\left(-\Omega^2\,\Delta_{max} + \omega^2\,\Delta_{max} - \frac{F}{M}\cos\phi \right) \sin(\Omega t - \phi)$$

$$+ \left(2\omega\zeta\Omega\,\Delta_{max} - \frac{F}{M}\sin\phi \right)\cos(\Omega t - \phi) = 0 \qquad (3.3)$$

This can be generally true only if the coefficients in front of the sine and cosine terms are each zero. That is

$$\Delta_{max} = \frac{F}{M(\omega^2 - \Omega^2)}\cos\phi$$

and

$$\Delta_{max} = \frac{F}{2\omega\Omega\zeta M}\sin\phi$$

which yields

$$\Delta_{max} = \frac{F/K}{\sqrt{\{[1 - (\Omega/\omega)^2]^2 + (2\zeta\Omega/\omega)^2\}}} \qquad (3.4)$$

and

$$\tan\phi = \frac{2\zeta\Omega/\omega}{1 - (\Omega/\omega)^2} \qquad (3.5)$$

Note that F/K is the 'static' deflection and usual practice is to define a dynamic amplification factor given by

$$DAF = \frac{\Delta_{max}}{(F/K)} = \frac{(K\,\Delta_{max})}{F}$$

that is

$$DAF = \frac{1}{\sqrt{\{[1 - (\Omega/\omega)^2]^2 + (2\zeta\Omega/\omega)^2\}}} \qquad (3.6)$$

The DAF depends on two dimensionless parameters: ζ, the fraction of critical damping, and Ω/ω, the frequency ratio.

85

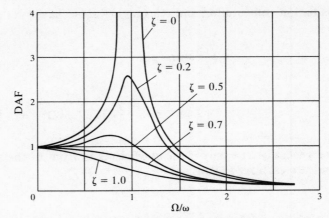

Figure 3.2 Peak steady-state response to $F \sin \Omega t$.

The form that this peak response takes – the response spectrum is another way of describing it – is given by the curves of Figure 3.2.

In point of fact when the system is started from rest the true peak response is generally slightly greater than that given by Equation 3.1, which, as we have said before, applies only after starting transients have died away. Qualitatively, this can be seen in Figure 3.1.

The idea of a response spectrum is a general one – it can be evaluated for any loading history for which explicit information is available. It is a useful design tool and, in fact, first came into common usage for the design of landing gear on aircraft: perhaps its most widespread use now in structural engineering relates to earthquake engineering, where response spectra may be calculated for different recorded seismic events.

3.1.1 Properties of the dynamic amplification factor

Physically speaking, the fraction of critical damping ζ is small in most structural applications. Bearing this in mind, we may consider the trends in the DAF as just the frequency ratio (Ω/ω) varies.

FREQUENCY RATIO $\Omega/\omega \rightarrow 0$

In this case DAF $\rightarrow 1$, for the spring–mass system is essentially rigid: the natural period of the system is much shorter than the period of the forcing function. The system's inertia force $M\ddot{\Delta}$ ($= -M\Omega^2 \Delta$, by definition) is vanishingly small and Equation 3.1 reduces to

$$\omega^2 \, \Delta \approx (F/M) \sin \Omega t$$

That is,

$$\Delta \approx (F/K) \sin \Omega t$$

which implies that DAF $\to$ 1. The level of damping has very little effect on this conclusion. In other words, the level of damping in very stiff structures need not be precisely determined when the loading has a frequency content in which low frequencies predominate.

FREQUENCY RATIO $\Omega/\omega \to \infty$

At the other extreme, Equation 3.6 indicates that DAF $\to 1/(\Omega/\omega)^2$. That is, the DAF becomes small and this conclusion is also essentially independent of the level of damping.

The spring–mass system is very flexible, the restoring force is relatively weak, and the nature of the solution may be found by solving (see Equation 3.1)

$$\ddot{\Delta} \approx (F/M) \sin \Omega t$$

namely

$$\Delta \approx -\frac{F/K}{(\Omega/\omega)^2} \sin \Omega t$$

which gives DAF $\approx 1/(\Omega/\omega)^2$. In this case, for given input, the actual peak dynamic displacement Δ_{max} is approximately constant, i.e. it is largely independent of system properties of stiffness and damping.

FREQUENCY RATIO $\Omega/\omega = 1$

For this particular value of the frequency ratio, the dynamic amplification factor is

$$\text{DAF} = 1/(2\zeta)$$

which is very large when ζ is small. Such behaviour is referred to as resonance (strictly speaking the peak in the DAF curve occurs to the left of $\Omega/\omega = 1$, but the effect is very slight when the damping is low).

In this central region of the frequency ratio, and particularly at resonance, the magnitude of the DAF is very strongly dependent on the level of damping. For example, going from $\zeta = 1\%$ to $\zeta = 2\%$ halves the response.

As was stated at the outset, steady-state response is of the form

$$\Delta = \Delta_{max} \sin(\Omega t - \phi)$$

which, by simple differentiation, implies that

$$\ddot{\Delta} + \Omega^2 \Delta = 0$$

But, at resonance, $\Omega \equiv \omega$, so the above is equivalent to

$$\ddot{\Delta} + \omega^2 \Delta = 0$$

If this is extracted from Equation 3.1 we are left with

$$2\zeta\omega\,\dot{\Delta} = (F/M)\sin\Omega t$$

or

$$2\zeta\Omega\dot{\Delta} = (F/M)\sin\Omega t$$

and so

$$\Delta(t) = -\frac{F/K}{2\zeta}\cos\Omega t$$

From this we recover the result that $\mathrm{DAF} = 1/(2\zeta)$ (see Problem 3.7).

Finally, we should comment in a semiquantitative way about a fundamental difference in response depending on which side of the resonant peak the structure happens to lie. For forcing of a given periodicity and magnitude, stiff structures show a decrease in response as their stiffness increases. This is indicated by the left-hand portion of Figure 3.2, and is compatible with what one expects from structural response in the realm of static loading: greater stiffness implies smaller response. However, flexible structures show an increase in response as their stiffness increases (see the right-hand portion of Figure 3.2). This trend has no counterpart in statics: it is purely dynamic in origin, being due to the reduction in phase angle between forcing and response at this end of the response spectrum.

3.1.2 Properties of the phase angle

Referring to Equation 3.5 we see that $\tan\phi \to 0$ when $\Omega/\omega \to 0$. That is, $\phi \to 0$ when $\Omega/\omega \to 0$, which is what one would expect because rigid-body motion must mean that response and forcing are in phase.

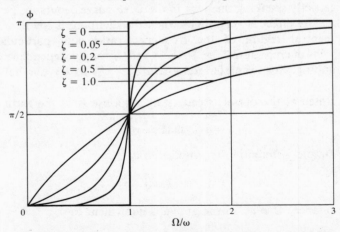

Figure 3.3 Phase in the steady-state response to $F\sin\Omega t$.

When $\Omega/\omega \to \infty$, $\tan \phi \to 0$ also, but $\tan \phi$ tends to zero from below, so that $\phi \to 180°$. Response and forcing are almost completely out of phase. It is not possible for system response to be more than $180°$ out of phase with the forcing: if that were indicated, the correct interpretation would be that the phase angle should be the complement of $360°$. In motion that is completely out of phase with the forcing, a positive peak in response corresponds to a negative in forcing.

At resonance $\tan \phi \to \infty$ and $\phi = 90°$: peak response occurs when the forcing is momentarily zero.

These trends are illustrated in Figure 3.3.

3.1.3 Energy relations

Under the assumption of steady-state response the energy input per cycle of vibration is exactly dissipated by viscous action: if this were not the case one could not claim that a true steady state had been reached − amplitudes would be either growing or shrinking with time.

At any point of any cycle the oscillator has kinetic energy of $\frac{1}{2}M\dot{\Delta}^2$ and potential energy of $\frac{1}{2}K\Delta^2$. One complete cycle later the kinetic energy is again $\frac{1}{2}M\dot{\Delta}^2$ and the potential energy is again $\frac{1}{2}K\Delta^2$. Therefore, over one cycle we have, from the energy equation

$$(\tfrac{1}{2}M\dot{\Delta}^2 - \tfrac{1}{2}M\dot{\Delta}^2) + (\tfrac{1}{2}K\Delta^2 - \tfrac{1}{2}K\Delta^2) = \oint F(t)\, d\Delta - \oint c\dot{\Delta}\, d\Delta$$

If we define the input energy over one cycle as

$$E_i = \oint F(t)\, d\Delta = \int_0^{2\pi/\Omega} F \sin(\Omega t)\dot{\Delta}\, dt$$

we arrive at

$$E_i = \pi F \Delta_{max} \sin \phi \tag{3.7}$$

after the necessary substitutions and integration have been performed.

Similarly, if the energy dissipated is defined as

$$E_d = \oint c\dot{\Delta}\, d\Delta = 2\zeta\omega M \int_0^{2\pi/\Omega} \dot{\Delta}^2\, dt$$

we obtain

$$E_d = 2\pi\zeta(\Omega/\omega)K\Delta_{max}^2 \tag{3.8}$$

Equating the two gives

$$\Delta_{max} = \frac{(F/K)\sin\phi}{2\zeta(\Omega/\omega)}$$

which is of identical form to one of the equations that led to the results given by Equations 3.4 and 3.5.

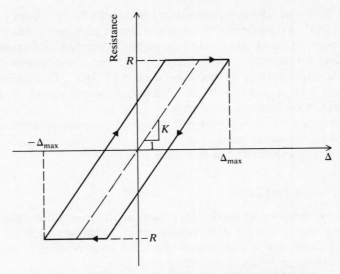

Figure 3.4 Hysteresis hoop for an elasto-plastic spring.

At resonance the energy input to the system is the maximum possible. However, we severely stress the assumption of linear response in those cases where, as at resonance, very large responses are predicted on account of assumed low values of damping. As has been mentioned before, damping levels are to some degree amplitude dependent and some amelioration of response necessarily attends that fact. In addition, there is a limit to the linearity that we assume in the equivalent spring. Softening of the spring due to local yielding causes the natural period to lengthen and the resonant peak is both shifted and diminished.

We can account for this in a semiquantitative way by again considering the energy balance, but with an allowance for plastic dissipation in addition to viscous damping. We may assume that the extent of yielding is not very pronounced and the force–displacement curve (the hysteresis loop) for the spring is as given in Figure 3.4.

In one cycle the loop is traversed once and the energy dissipated is

$$2R\,[\,2(\Delta_{max} - R/K)\,] = 4R(\Delta_{max} - R/K)$$

where R is the upper limit on the resistance available in the spring. Thus, as in Chapter 2, R/K is the displacement at which the elastic limit is reached.

Let us suppose that Equation 3.7 holds and that peak steady-state elasto-plastic response is defined by $\sin \Omega = 1$: this is certainly plausible. The input energy is then

$$E_i = \pi F \Delta_{max}$$

and the dissipated energy comprises two parts, as follows

$$E_d = 2\pi\zeta K\Delta_{max}^2 + 4R(\Delta_{max} - R/K)$$

We can define equivalent viscous damping in the present context as given by

$$E_d = 2\pi\zeta_{eq}K\Delta_{max}^2$$

and further define $\mu = \Delta_{max}/(R/K)$: this is the ductility factor introduced in the last chapter. After some rearrangement we obtain

$$\zeta_{eq} = \zeta + \frac{2}{\pi}\left(\frac{1}{\mu} - \frac{1}{\mu^2}\right) \qquad \mu \geqslant 1 \qquad (3.9)$$

For example, suppose $\mu = 1.05$ – that is, some slight yielding is taking place. The equivalent viscous damping is

$$\zeta_{eq} = \zeta + 0.029$$

If $\zeta = 0.02$, say,

$$\zeta_{eq} = 0.049$$

and the peak resonant response is only about 40% of that based on the assumption of purely elastic response.

Clearly, very little yielding is necessary to reduce resonant peaks significantly.

More generally, in those situations where the yield stresses may just be reached during dynamic response, we could justifiably use larger values of ζ and imagine that the structure is still responding elastically. In particular, as the above numbers imply, the damping values given earlier in Table 2.2 could be arbitrarily increased by a factor of 2 to allow for this eventuality. This is consistent with the recommendations of authorities such as Newmark and Hall (1982).

Example 3.1 Dynamic response characteristics of a tension-leg platform

A partial general arrangement of a tension-leg platform is shown in Figure 3.5. The purpose of this exercise is to estimate the forces and displacements induced by specified wave action. The wave forcing will be taken as being represented by a simple sinusoidal term and we shall concentrate on just two aspects: heave (i.e. vertical motion) and surge (i.e. horizontal motion in direction of wave attack) responses. Movement due to current loading is here ignored.

The platform floats because the deck load is balanced by the buoyancy provided by the six legs and six submerged pontoons. The system is stayed by four vertical tension legs, which consist of clusters of high-strength circular hollow sections. Flexible joints are provided at the mudline and the soffit of the large-diameter

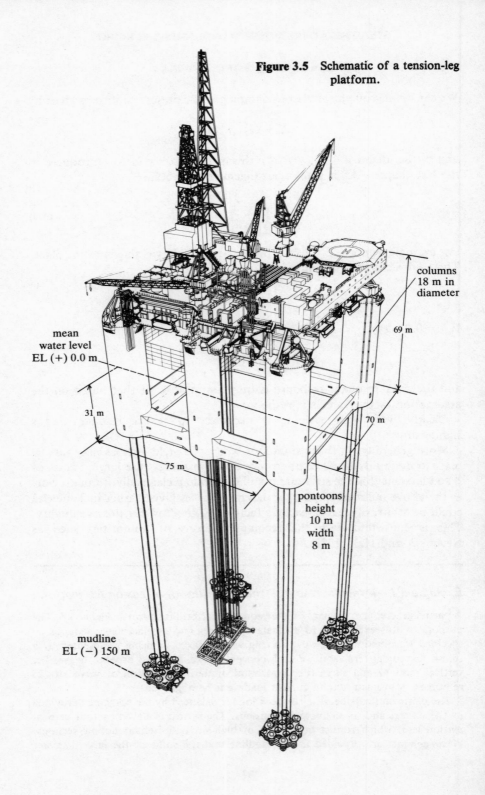

Figure 3.5 Schematic of a tension-leg platform.

columns
18 m in
diameter

69 m

mean
water level
EL (+) 0.0 m

31 m

70 m

75 m

pontoons
height
10 m
width
8 m

mudline
EL (−) 150 m

buoyant legs: this minimises the local effects of bending and the structure acts as a massive inverted pendulum in responding to horizontal forcing. The horizontal stiffness is purely gravitational in origin because it is due to leg tension, which is itself generated by mobilising additional buoyancy (over and above that due to self-weight) from the large-diameter legs. On the other hand, vertical stiffness is purely elastic in origin, being due to the axial extensions of the tension legs. As might be expected, the vertical stiffness is several orders of magnitude greater than the lateral stiffness. Particular details are:

mean water level 150 m
mudline 0
total weight 500 000 kN
displacement at MWL 620 000 kN
total present tension (for four legs) 120 000 kN
plan dimensions 75 m × 70 m
buoyant legs (six) 18 m O.D.
pontoons (six between legs, height × width) 10 m × 8 m
tension legs (each leg) 4 × 250 O.D. × 100 wall thickness
submergence of platform (this can be calculated from the above information) 31 m
length of tension legs $150 - 31 = 119$ m
specific weight of sea water 10 kN m^{-3}

FREE VIBRATION CHARACTERISTICS

Surge motion
The equation of motion is

$$M_s \ddot{\Delta}_s + K_s \Delta_s = 0$$

where $K_s = T/L = 120\,000 \times 9.81/119 = 1000$ kN m^{-1}. The mass M_s consists of two parts: the first is simply the mass of the platform (50 000 kN m s^{-2}), and the second is the added mass due to the fact that the buoyant legs and pontoons accelerate the ambient water with which they are in contact. The theoretical result for circular cylinders of infinite length is that the added mass mobilised when they are accelerated perpendicular to their axes of symmetry is equal to the mass of fluid displaced. Thus the buoyant legs will be taken as cylinders of diameter 18 m, as given, while the pontoons will be taken to be cylinders of diameter 10 m for present purposes. Including various shielding effects the added mass is approximately

$$2 \times 39 \times \frac{\pi \times 10^2}{4} + 6 \times 31 \times \frac{\pi \times 18^2}{4} = 53\,500 \text{ kN s}^2\text{m}^{-1}$$

This yields a total mass in surge of

$$M_s = 50\,000 + 53\,500$$

$$= 103\,500 \text{ kN s}^2\text{m}^{-1}$$

Thus the period of surge is

$$T_s = 2\pi \sqrt{(M_s/K_s)} = 64 \text{ s}$$

In surge, then, the tension-leg platform is a long-period structure by any definition.

93

Heave motion

The equation of motion is

$$M_h \ddot{\Delta}_h + K_h \Delta_h = 0$$

where, in this case, K_h is elastic in origin. Each of the four tension legs consists of a cluster of four high-tensile steel tubes. Hence, the total axial stiffness is

$$K_h = \frac{200 \times 4 \times \frac{1}{4}\pi(250^2 - 100^2) \times 4}{119}$$

$$= 1.11 \times 10^6 \ \mathrm{kN\,m^{-1}}$$

The added mass of the platform is less in the heave mode and is estimated to be about $15\,000 \ \mathrm{kN\,s^2\,m^{-1}}$. Hence the total heave mass is

$$M_h = 65\,000 \ \mathrm{kN\,s^2\,m^{-1}}$$

The period of heave motion is

$$T_h = 2\pi\sqrt{(M_h/K_h)} = 1.5 \ \mathrm{s}$$

ESTIMATE OF RESPONSE TO THE OPERATING WAVE

The peak operating wave is defined as the largest wave for which platform equipment is designed to remain operational. If such a wave is forecast then, in all probability, the platform would be shut down and personnel brought ashore. For comparative puposes we shall consider steady-state response to a train of operating waves of period 11 s, and crest-to-trough height of 12 m. The wavelength is about 190 m, which is not large compared to the plan dimensions of the platform, so wave phase effects are important even though we choose to ignore them here.

For heave motion the applied force may be based on the variation in volume displaced by the buoyant legs for a rise and fall each of 6 m. This force is of magnitude

$$F_h = 6 \times \frac{\pi \times 18^2}{4} \times 6 \times 9.81$$

$$= 9.0 \times 10^4 \ \mathrm{kN}$$

and has a period of 11 s. From Equation 3.6 we have,

$$\mathrm{DAF} = \frac{1}{|\,1 - (\Omega/\omega)^2\,|}$$

if damping is ignored. However $\Omega/\omega = 1.5/11 = 0.136$, which is small and so damping does not play a significant part in the proceedings. Specifically

$$\mathrm{DAF} = \frac{1}{|\,1 - 0.136^2\,|} = 1.02$$

Thus, as expected, there is very little dynamic amplification: the structure is so stiff in heave motion that it responds statically to the much longer-period wave forcing. The axial extension of the tension legs under this wave loading is

$$\Delta_h = DAF \times F_h/K_h$$

$$= 1.02 \times \frac{9 \times 10^4}{1.11 \times 10^6}$$

$$= 0.083 \text{ m}$$

The tension in the legs changes periodically in response to heave motion. The preset tension is about 80% lost when the wave trough goes by. By the same token, it almost doubles when a crest passes by. Obviously, this type of behaviour is of importance in assessment of fatigue life.

It is also of importance in assessing the surge response. Since the heave motion changes the leg tension, and the lateral stiffness of the platform is dependent on that tension, it is not a straightforward matter to calculate response in surge. Having said that, we shall ignore this complication for the time being (it will be picked up again in Chapter 4) and assume that the leg tension is constant with time.

If the platform was rigid the surge wave loading, F_s, would be about 6×10^4 kN. The frequency ratio is $\Omega/\omega = 64/11 = 5.82$. Because the platform is so flexible in surge, damping is largely irrelevant and the dynamic amplification factor is

$$DAF \approx \frac{1}{|1 - 5.82^2|}$$

$$\approx 0.030$$

The resisting force in the tension legs is thus quite low because the inertia of the system is relatively large. Nevertheless, the drift under the wave loading is

$$\Delta_s = DAF \times F_s/K_s$$

$$= 0.030 \times 6 \times 10^4/990$$

$$= 1.8 \text{ m}$$

which is large by the normal criteria of structural engineering. Of course, this must be regarded as an estimate only for there are several complicating factors to do with the nature of the loading and phase effects due to the sheer size in plan of the platform. For further information on waveloading and related matters see Brebbia & Walker (1979) and Newman (1978).

3.2 Transient response to sinusoidal forcing

The dynamic amplification factor given by Equation 3.6 applies only after all transients due to sudden imposition of the loading have died out. In general, in the case where the oscillator starts from rest under $F \sin \Omega t$, the peak response may occur early on and be greater than that given by Equation 3.6. The general solution with zero initial conditions is

$$\frac{\Delta(t)}{\text{DAF} \times F/K} = e^{-\omega \zeta t} \left(\sin \phi \cos \omega_d t + \frac{\zeta \sin \phi - (\Omega/\omega) \cos \phi}{\sqrt{(1-\zeta^2)}} \sin \omega_d t \right)$$
$$+ \sin(\Omega t - \phi) \qquad (3.10)$$

Previously, Figure 3.1 has been used to depict how response settles down to steady state when the system natural frequency is high. The transients soon die out in this case because their periodicity is that of the system and in such instances damping takes its toll faster. However, when the system natural frequency is low, many more cycles are generally required before it can be claimed that a steady state has been reached. In absolute terms a good deal more time has to go by before the steady state is reached – it will be recalled that this point has been made before in a slightly different context in Chapter 2.

For zero damping the dynamic amplification factor for the steady state is (look back at Eqn 3.6)

$$\text{DAF} = \frac{1}{|1 - (\Omega/\omega)^2|}$$

Transient response when damping is ignored (thus the phase angle ϕ may be taken as zero) is of the form

$$\frac{\Delta(t)}{F/K} = \frac{1}{1 - (\Omega/\omega)^2} \left(\sin \Omega t - \frac{\Omega}{\omega} \sin \omega t \right)$$

The largest possible value of this quantity would occur when $\sin \Omega t = 1$ and $\sin \omega t = -1$, leading to the maximum possible value of the dynamic amplification for transient response to sinusoidal loading to

$$\text{DAF} = \frac{1 + (\Omega/\omega)}{1 - (\Omega/\omega)^2} = \frac{1}{|1 - (\Omega/\omega)|} \qquad (3.11)$$

Since it may well be that peak transient response occurs too soon for damping to have had much effect, Equation 3.11 stands as a useful design formula when transient response to sinusoidal loading is under consideration. Again, the comment may be made that Figure 3.1 provides a useful qualitative illustration of this transient phenomenon. Equation 3.11 could be usefully applied to Example 3.1 (the tension-leg platform) in order to estimate the severest response that an 'isolated' wave could engender.

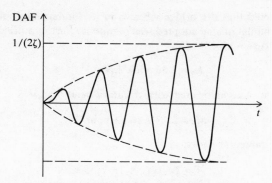

Figure 3.6 Growth of resonant response.

The other case of interest, for which a sketch is justified, concerns the growth of resonant response from a sudden start. The curve in Figure 3.6 is for $\omega = \Omega$ and small damping. Its equation is

$$\frac{\Delta(t)}{\text{DAF} \times F/K} \approx -(1 - e^{-\zeta\Omega t}) \cos \Omega t$$

The coefficient $(1 - e^{-\zeta\Omega t})$ eventually 'reaches' unity and this signals the onset of a resonant response, namely

$$\frac{\Delta(t)}{F/K} = \frac{\cos \Omega t}{2\zeta} \tag{3.12}$$

The point may thus be made that a resonant response takes some time to develop even when the forcing is of precisely the right frequency. If there are only a few cycles at the resonant frequency, the build-up of response may not be significant at all. The spin cycle in a washing machine affords a good example of a structure that passes through resonance both in spinning up and slowing down: in the more expensive machines, the process is well managed and vibration levels are tolerably low.

Example 3.2   Dynamic response calculations for a bridge truss under idealised vehicular loading

The purpose of this example is to illustrate transient response to 'quasi-sinusoidal' loading using the bridge truss of Problem 1.2. To this end we consider the midspan deflection of a 40 m span steel-truss bridge under a travelling point load (representing vehicular traffic) as the speed of crossing varies. However, the dynamic characteristics of vehicle itself are ignored, as are surface roughness effects: more general discussion can be found in Biggs (1964), for example.

It will be assumed that the bridge vibrates in its fundamental mode: it will be recalled that the modal profile adopted was parabolic. Furthermore, the equivalent stiffness of one truss was

$$K_e = 3.33 \times 10^3 \text{ kN m}^{-1}$$

and the equivalent mass associated with tributary dead load was

$$M_e = 2.17 \times 10 \text{ kN s}^2 \text{ m}^{-1}$$

Hence the fundamental period is

$$T = 2\pi \sqrt{(M_e/K_e)} = 0.51 \text{ s}$$

Consider, now, the equation of motion for undamped response to a point load P which enters the bridge from one side and travels across with velocity V (see Fig. 3.7).

The profile of deflection in the span at any instant is

$$\Delta(x, t) = 4 \left(\frac{x}{L}\right) \left(1 - \frac{x}{L}\right) \Delta_{\text{dyn}}(t)$$

The expression for the equivalent loading is

$$P_e = \int_0^L 4 \left(\frac{x}{L}\right) \left(1 - \frac{x}{L}\right) \rho(x) \, \mathrm{d}x$$

where $\rho(x)$ is a distributed loading. In our case $\rho(x)$ is zero everywhere except at the position of the point load where $\rho(x) \, \mathrm{d}x = P$. Hence the integral is

$$P_e = 4 \left(\frac{x}{L}\right) \left(1 - \frac{x}{L}\right) P$$

However, in the present context

$$x = Vt$$

for $t \leqslant L/V$ and so the loading term has the form

$$P_e(t) = 4 \left(\frac{Vt}{L}\right) \left(1 - \frac{Vt}{L}\right) P$$

and this holds only while the vehicle is on the span.

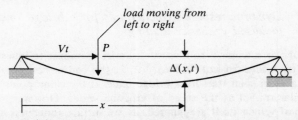

Figure 3.7 Definition diagram for bridge and co-ordinate system adopted.

Before proceeding further we note the similarity of the function $4(Vt/L)(1 - Vt/L)$ to the sinusoidal function $\sin(\pi Vt/L)$ for the timespan that the vehicle is on the bridge. Thus the response we will obtain for the midspan deflection of the bridge as the vehicle goes by will be given by an equation quite similar to Equation 3.10. Here, however, we will ignore damping.

The equation of motion is

$$M_e \ddot{\Delta}_{dyn} + K_e \, \Delta_{dyn} = 4 \, \frac{Vt}{L} \left(1 - \frac{Vt}{L} \right) P \qquad 0 \leqslant t \leqslant L/V$$

which may be rearranged to

$$\ddot{\Delta}_{dyn} + \omega^2 \, \Delta_{dyn} = 4 \, \frac{P}{K_e} \, \omega^2 \, \frac{Vt}{L} \left(1 - \frac{Vt}{L} \right)$$

The quantity P/K_e is the static midspan deflection when the load is at midspan and we may write it as Δ_{st}. Thus

$$\ddot{\Delta}_{dyn} + \omega^2 \, \Delta_{dyn} = \omega^2 \, \Delta_{st} 4 \, \frac{Vt}{L} \left(1 - \frac{Vt}{L} \right)$$

The complementary function is straightforward, being, as always,

$$A \cos \omega t + B \sin \omega t$$

The particular integral may be seen to be

$$\Delta_{st} 4 \, \frac{Vt}{L} \left(1 - \frac{Vt}{L} \right) + 8 \, \Delta_{st} \, \frac{V^2}{\omega^2 L^2}$$

and the specific solution, which also satisfies initial conditions of zero displacement and zero velocity, is

$$\frac{\Delta_{dyn}(t)}{\Delta_{st}} = 8 \, \frac{V^2}{\omega^2 L^2} \, (1 - \cos \omega t) - 4 \, \frac{V}{\omega L} \sin \omega t + 4 \, \frac{Vt}{L} \left(1 - \frac{Vt}{L} \right)$$

The solution holds only while the load is on the bridge.

In order to get some feel for the result we use the following data to obtain the peak value of the midspan deflection. This again raises the concept of a dynamic amplification factor, more commonly referred to as the impact factor. Let us assume that the velocity of the vehicle is 20 m s^{-1}: it is on the bridge for $40/20 = 2$ s. The natural frequency is 12.6 rad s^{-1}.

Figure 3.8 shows the midspan deflection response as the load moves across. Not surprisingly the largest dynamic effect occurs when the load is close to midspan. The fact that there is a dynamic effect is, of course, because the load is travelling a curved path owing, in the first instance, to the deflection due to vehicle weight. This slight curvature generates a centrifugal acceleration. Note that the period of the bridge is 25% of the crossing time and this is evident in Figure 3.8.

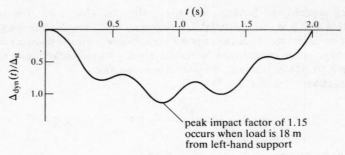

peak impact factor of 1.15
occurs when load is 18 m
from left-hand support

Figure 3.8

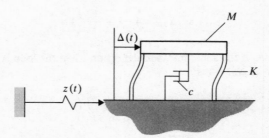

Figure 3.9 Definition diagram for base excitation.

3.3 Response to ground motion

We consider here steady-state response to sinusoidal forcing which comes to the structure through its foundation. The most obvious application of this approach concerns earthquake response, but there are a whole host of others, including equipment and secondary structures mounted in or on other structures. Ships and aircraft are also systems that are subject to 'support' motions.

The diagram in Figure 3.9 shows a base-pad undergoing absolute movement

$$z(t) = Z \sin \Omega t$$

while the response displacement $\Delta(t)$ is regarded as being measured relative to the base. The absolute movement of the mass is thus $\Delta(t) + z(t)$; the absolute acceleration is therefore $\ddot{\Delta}(t) + \ddot{z}(t)$ and, from Newton's second law, we have

$$M(\ddot{\Delta} + \ddot{z}) + c\dot{\Delta} + K\Delta = 0$$

or

$$\ddot{\Delta} + 2\zeta\omega\,\dot{\Delta} + \omega^2\,\Delta = -\ddot{z}(t) \tag{3.13}$$

In the particular case treated herein

$$\ddot{z} = -\Omega^2 Z \sin \Omega t$$

and so we have

$$\ddot{\Delta} + 2\zeta\omega\dot{\Delta} + \omega^2 \Delta = \Omega^2 Z \sin \Omega t$$

This is identical with Equation 3.1 provided we define

$$F/M = \omega^2 F/K \equiv \Omega^2 Z$$

Hence Equation 3.4 may be rewritten as

$$\Delta_{max} = \frac{(\Omega/\omega)^2 Z}{\sqrt{\{[1-(\Omega/\omega)^2]^2 + (2\zeta\Omega/\omega)^2\}}}$$

or

$$\frac{\Delta_{max}}{Z} = \frac{(\Omega/\omega)^2}{\sqrt{\{[1-(\Omega/\omega)^2]+(2\zeta\Omega/\omega)^2\}}} \qquad (3.14)$$

and we have an expression for the peak relative displacement. In Figure 3.10 plots of Equation 3.14 are shown for varying values of the frequency ratio and damping. The behaviour at each end of the frequency spectrum is of most direct interest.

WHEN Ω/ω IS SMALL

In this case we have

$$\Delta_{max}/Z \approx (\Omega/\omega)^2 \qquad (3.15)$$

This corresponds to rigid-body response and the phase angle is small.

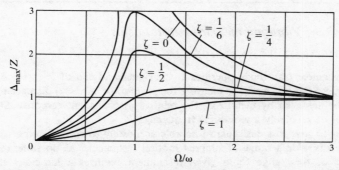

Figure 3.10 Peak relative displacement response for sinusoidal base excitation.

101

WHEN Ω/ω IS LARGE

Here we obtain

$$\Delta_{max}/Z \approx 1 \qquad (3.16)$$

Because response is almost completely out of phase with forcing, the absolute displacement of the mass, viz. $\Delta(t) + z(t)$, is close to zero. That is, because

$$\Delta(t) \approx -Z \sin \Omega t$$

then

$$\Delta(t) + z(t) \approx 0$$

We have here an explanation of why tall buildings attract little loading during an earthquake, for the period of the structure is typically much longer than the predominant period in the earthquake ground motion. In many respects tall buildings are the safest places to be during an earthquake, a point which is not easy to get across to the general public (see Example 4.2).

The surge response calculations for the tension-leg platform conducted in Example 3.1 well illustrate this point also. Because of its flexibility, the value of the actual dynamic load was only 3% of the wave load capable of being applied to the platform if it had been rigidly founded on the sea floor. That sort of figure is typical of very tall buildings responding to earthquake loading and almost invariably means that wind loading governs.

The examples that follow introduce two types of vibration-measuring instrument, which operate at different ends of the response spectrum (as given by Eqn 3.14), and a method is developed for the calculation of damping from resonant response curves.

Example 3.3 Vibration-measuring instruments

DISPLACEMENT TRANSDUCER

The displacement transducer operates at the right-hand end of Figure 3.10, where Ω/ω is large. So if one wants to measure the ground (or support) displacement $z(t)$, this can be achieved by measuring the relative displacement response $\Delta(t)$ in an instrument of relatively low natural frequency.

Needless to say, it is desirable to be able accurately to record support displacements over as wide a range of support motion frequencies as possible: to tap the lower end of this range for a given instrument requires a judicious choice of damping.

Mathematically speaking we require (recall Eqn 3.14)

$$\sqrt{\{[1 - (\Omega/\omega)^2]^2 + (2\zeta\Omega/\omega)^2\}} \approx (\Omega/\omega)^2$$

for (Ω/ω) as small as possible. Squaring both sides and ordering terms gives

$$(\Omega/\omega)^4 - 2(1 - 2\zeta^2)(\Omega/\omega)^2 + 1 \approx (\Omega/\omega)^4$$

Left-hand side and right-hand side are broadly in conformity if the damping is chosen such that

$$\zeta = 1/\sqrt{2} = 0.71$$

Thus, the displacement transducer requires high (and stable) damping in order to be successful over a wide range of input frequencies. In modern electronic instruments this damping is based on the eddy current principle. The practical range of damping values is $\zeta = 0.6-1.0$. The value $\zeta = 0.6$ is quite common.

Displacement transducers come in varying sizes depending on the magnitude of the support displacement that is to be measured. They are not a practicable solution where displacements of the order of 10–20 cm, or more (typical, say, of strong earthquake response levels in the upper storeys of a high-rise building), are to be recorded, because the instrument is quite simply unwieldy.

In such cases, at least, it is better to record support acceleration and, if necessary, integrate the record to obtain displacements. The instrument that records acceleration is the accelerometer.

ACCELEROMETER

Accelerometers work at the left-hand end of the response spectrum (see Fig. 3.10), where Ω/ω is small. This implies an instrument of small mass and a stiff spring, as opposed to the displacement transducer which often has a 'soft' system. In the region of interest

$$\Delta(t) \approx \frac{\Omega^2 Z}{\omega^2} \sin \Omega t$$

and, since ω^2 is an instrument constant,

$$\Delta \sim \Omega^2 z \sim \ddot{z}$$

Hence, by measuring the relative displacement between instrument mass and baseplate we obtain a record of the actual support acceleration.

Mathematically speaking, we require the parabolic relationship

$$\Delta_{max}/Z \sim (\Omega/\omega)^2$$

to exist for as wide a range of forcing frequencies as possible. Since the instrument's natural frequency is usually relatively high to start with, we are not concerned with the accurate measurement of low-frequency phenomena (like response of a tall building) because this is automatically guaranteed. To extend the range in the other direction we require that

$$\sqrt{\{[1-(\Omega/\omega)^2]^2+(2\zeta\Omega/\omega)^2\}} \approx 1$$

which again boils down to

$$\zeta = 1/\sqrt{2} = 0.71$$

and it is again true that a value $\zeta \approx 0.6$ seems to be best.

In the context of the accelerometer the term 'amplitude distortion' is the amount by which the term

$$\sqrt{\{[1-(\Omega/\omega)^2]^2+(2\zeta\Omega/\omega)^2\}}$$

differs from unity, for a given instrument measuring forcing frequencies of a stated value.

The phase distortion present in the instrument is another quantity of direct interest. Recorded motions are rarely the pure sinusoids on which the present theory is based. The measurement of earthquake ground motion, for example, has more in common with measuring a random waveform than anything else. Such a waveform can be thought of as composed of several component waves, however, and it is obviously important that the instrument combines these waves correctly. The phase relationships must be preserved.

For purposes of demonstration, suppose that the ground motion is given by

$$z(t) = Z_1 \sin \Omega_1 t + Z_2 \sin \Omega_2 t$$

The instrument response is

$$\Delta(t) \approx \frac{1}{\omega^2} [\Omega_1^2 Z_1 \sin(\Omega_1 t - \phi_1) + \Omega_2^2 Z_2 \sin(\Omega_2 t - \phi_2)]$$

$$\approx \frac{1}{\omega^2} \left\{ \Omega_1^2 Z_1 \sin\left[\Omega_1\left(t-\frac{\phi_1}{\Omega_1}\right)\right] + \Omega_2^2 Z_2 \sin\left[\Omega_2\left(t-\frac{\phi_2}{\Omega_2}\right)\right] \right\}$$

To maintain relative phase between input and output we require that

$$\frac{\phi_1}{\Omega_1} = \frac{\phi_2}{\Omega_2} = \frac{\phi}{\Omega} = \text{constant}$$

However, when Ω/ω is small the phase angle is given by

$$\tan \phi \approx \phi \approx 2\zeta\Omega/\omega$$

which is of the required form, namely $\phi/\Omega \approx$ constant.

104

One may again go through an argument to establish a value of damping such that phase distortion is minimal over a wide range of frequencies. This leads to a value of $\zeta = \pi/4 = 0.785$: fortunately this is compatible with the requirements for nominal amplitude distortion. Look at Figure 3.3 and notice that a value of $\zeta = 0.785$ comes close to possessing the desired characteristics of straight-line behaviour implied by $\phi = \text{constant} \times \Omega$.

Phase distortion is sometimes measured as the difference (in radians) between ϕ and the quantity $\frac{1}{2}\pi\Omega/\omega$.

A wide range of vibration-measuring instruments are available, so that frequency limits and various limits on displacement or acceleration response can generally be accommodated: it is very much a case of horses for courses.

Example 3.4   Calculation of damping from resonant response curves

In conducting an experimental search for the natural frequencies and damping levels in a building it is common to use some sort of variable-drive eccentric mass excitation. By varying the frequency of the eccentric mass exciter in discrete steps, and recording the resulting steady-state response, one obtains a series of resonant peaks and adjacent troughs in which each peak signals close proximity to a system natural frequency. An isolated peak, depicting the fundamental natural frequency of the system, would look much as in Figure 3.11a.

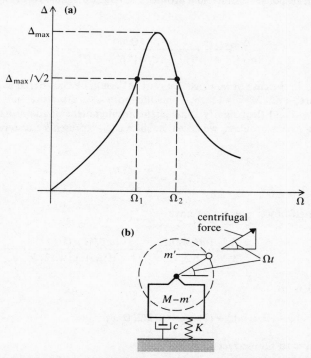

Figure 3.11   (a) Steady-state response as a function of forcing frequency. (b) Eccentric mass excitation.

105

The peak amplitude, Δ_{max}, occurs at the natural frequency, ω, but it is convenient to label two further points, Ω_1 and Ω_2, at which the response is $\Delta_{max}/\sqrt{2}$: these are the 'half-power' points. By finding Ω_1 and Ω_2 in this way it is possible to calculate the fraction of critical damping, ζ, without having recourse to such imponderables as loading level and system mass.

In Figure 3.11b we have a model of a simple system subjected to eccentric mass excitation. The angle swept out by the rotating mass m', of eccentricity ϵ, is Ωt and so the vertical component of the centrifugal force is

$$m' \Omega^2 \epsilon \sin \Omega t$$

The full mass, $(M - m') + m'$, has an acceleration $\ddot{\Delta}$, so that the equation of motion reads

$$M\ddot{\Delta} + c\dot{\Delta} + K\Delta = m' \Omega^2 \epsilon \sin \Omega t$$

or

$$\ddot{\Delta} + 2\zeta\omega\dot{\Delta} + \omega^2 \Delta = (m'/M)\epsilon\Omega^2 \sin \Omega t$$

which is of precisely the same form as the equation of motion governing relative displacement response to sinusoidal ground acceleration. Thus, from Equation 3.14 we have

$$\frac{\Delta_{max}}{(m'/M)\epsilon} = \frac{(\Omega/\omega)^2}{\sqrt{\{[1 - (\Omega/\omega)^2]^2 + (2\zeta\Omega/\omega)^2\}}}$$

Incidentally, in the case of an electric motor m' could be considered as the mass of the armature, and M would then include both the armature and stator: the eccentricity ϵ could then signify some slight misalignment of the armature.

Returning to our problem, we note that since Δ_{max} represents a resonant peak, we have

$$\Delta_{max} = \frac{(m'/M)\epsilon}{2\zeta}$$

So, at the half-power points, we have

$$\frac{\Delta_{max}}{\sqrt{2}} = \frac{(m'/M)\epsilon}{2\sqrt{2}\zeta} = \frac{(m'/M)\epsilon(\Omega/\omega)^2}{\sqrt{\{[1 - (\Omega/\omega)^2]^2 + (2\zeta\Omega/\omega)^2\}}}$$

This gives rise to

$$[1 - (\Omega/\omega)^2]^2 + (2\zeta\Omega/\omega)^2 = 2(2\zeta\Omega/\omega)^2$$

which reduces to the perfect square

$$[1 - (\Omega/\omega)^2]^2 = (2\zeta\Omega/\omega)^2$$

In taking the square root we can have both positive and negative signs:

$$1 - (\Omega/\omega)^2 = \pm 2\zeta\Omega/\omega$$

In the case of a positive sign we have

$$(\Omega/\omega)^2 + 2\zeta(\Omega/\omega) - 1 = 0$$

Since there is just one change of sign in the coefficients of this quadratic there is, from Descartes' rule of signs, just one positive real root. This is

$$\Omega/\omega = -\zeta + \sqrt{(1 + \zeta^2)}$$

But since ζ is assumed small we have

$$\Omega_1/\omega \simeq 1 - \zeta$$

Similarly, when the negative sign is taken we have

$$(\Omega/\omega)^2 - 2\zeta(\Omega/\omega) - 1 = 0$$

which, again, has only one change of sign in its coefficients and the root we require is

$$\Omega_2/\omega \simeq 1 + \zeta$$

Consequently, by rearrangement and substitution we arrive at the formula

$$\zeta \simeq \left(\frac{\Omega_2 - \Omega_1}{\Omega_1 + \Omega_2}\right)$$

and we have a straightforward way of determining system damping from experimentally determined response curves (see Problem 3.3).

3.4 Vibration transmission and isolation

Vibration due to machinery with rotating parts, for example, will be transmitted to the supports. Depending on their level these received vibrations may be objectionable. So it is in the case of attempting to isolate sensitive equipment from surroundings in which ambient vibrations are present. In both cases suitably designed sprung supports can go a long way to solving the problem but, as usual, a compromise must be reached between desirable system flexibility and simple functional requirements.

In this section the theory will be presented for transmissibility only: the same conclusions hold for isolation. A system exhibiting steady-state response to $F \sin \Omega t$ transmits a force

$$F_t = c\dot{\Delta} + K\Delta \tag{3.17}$$

to its foundation, where

$$\Delta(t) = \frac{(F/K)\sin(\Omega t - \phi)}{\sqrt{\{[1 - (\Omega/\omega)^2]^2 + (2\zeta\Omega/\omega)^2\}}}$$

and

$$\dot{\Delta}(t) = \frac{(\omega F/K)\cos(\Omega t - \phi)}{\sqrt{\{[1 - (\Omega/\omega)^2]^2 + (2\zeta\Omega/\omega)^2\}}}$$

After some rearrangement one obtains

$$\frac{F_t}{F} = \sqrt{\left(\frac{1 + (2\zeta\Omega/\omega)^2}{[1 - (\Omega/\omega)^2]^2 + (2\zeta\Omega/\omega)^2}\right)} \sin(\Omega t - \phi + \gamma)$$

where $\tan \gamma = 2\zeta\Omega/\omega$, in which γ is a further phase angle. The transmissibility is

$$T_r = \sqrt{\left(\frac{1 + (2\zeta\Omega/\omega)^2}{[1 - (\Omega/\omega^2]^2 + (2\zeta\Omega/\omega)^2}\right)}$$

which is plotted in Figure 3.12.

It is easily shown that:

$$T_r \geqslant 1 \qquad \text{when} \qquad 0 \leqslant \Omega/\omega \leqslant \sqrt{2}$$

and that:

$$T_r < 1 \qquad \text{when} \qquad \Omega/\omega > \sqrt{2}$$

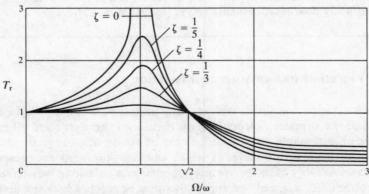

Figure 3.12 Plots of vibration transmissibility.

This latter case delineates the region of interest. For a really flexible mounting

$$T_r \to \frac{2\zeta\Omega/\omega}{(\Omega/\omega)^2} \to \frac{2\zeta}{(\Omega/\omega)}$$

We see that a further requirement is that damping be as low as possible, otherwise the transmitted force will be unnecessarily high. The book by Arya *et al.* (1979) on the design of structures and foundations for vibrating machines contains much practical information on this important topic.

Problems

3.1 The bridge vibration exercise conducted as Example 3.2 is in the form such that further information can be extracted. First, calculate the impact factors for crossing speeds of $10\,\mathrm{m\,s^{-1}}$ and $30\,\mathrm{m\,s^{-1}}$ and compare the results. Secondly, discuss how the theory presented can be adapted to cover the more realistic case of a pair of wheel loads crossing in tandem.

3.2 A drop hammer is found to transmit large shock loadings to the surrounding ground. To reduce this the machine is mounted on springs, as shown in Figure 3.13.

To prevent undue vibration after impact, damping is introduced. The constants of the system are: $W_1 = 10\,\mathrm{kN}$, $W_2 = 150\,\mathrm{kN}$, $h = 2.5\,\mathrm{m}$, $K = 4 \times 10^3\,\mathrm{kN\,m^{-1}}$ (total) and $\zeta = 0.05$. The weight W_1 falls through the height h and does not rebound. Find the peak force transmitted to the ground as a result of the impact. What is the value of the transmitted force three complete cycles after the peak occurs?

3.3 Figure 3.14 shows the results of two forced vibration tests for an eight-storey building. The output is response acceleration as depicted on the ordinate.

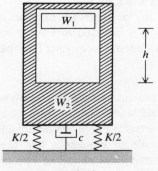

Figure 3.13

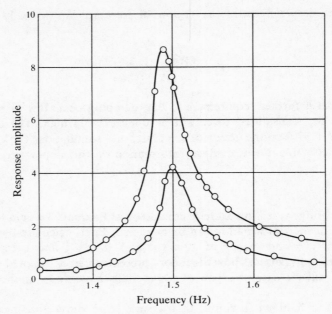

Figure 3.14

First, estimate the natural period of the system. Likewise, estimate the damping levels indicated in each test, and discuss the discrepancy.

3.4 Repeat the analysis of the tension-leg platform considered in Example 3.1 for the case when the water depth is 300 m. It may be assumed that platform characteristics are unchanged, as is wave loading.

3.5 The Philips PR 9266 displacement transducer is of diameter 58 mm and height 101 mm. Its total mass is about 490 g, although the vibrating mass is only 20 g. Its natural frequency (in the absence of damping) is 14 Hz. The damping is $\zeta = 0.6$.

The instrument is to be used to obtain displacement levels in an instrument panel, which is believed to have a resonant frequency around 40 Hz. For this value of forcing frequency find the amplitude distortion that the instrument will display.

3.6 This problem concerns a dynamic analysis of a very tall circular reinforced concrete wind shield for the flues of a coal-fired power station. The exercise, which is lengthy, involves determining the natural frequency of the fundamental mode of vibration, and then obtaining the dynamic amplification factors for vortex shedding loading under a range of wind velocities. The structure is shown schematically in Figure 3.15.

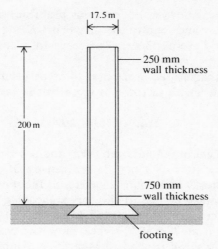

Figure 3.15

The outer diameter of the shield is constant at 17.5 m, the wall thickness varies continuously from 750 mm at the base to 250 mm at the top, and the height is 200 m. In addition, the concrete compressive strength varies from 35 MPa at the base to 25 MPa at the top, so that Young's modulus shows a significant variation over the height. The total weight (shield plus internal flues and ancillary structure) is estimated to vary from 1500 $kN\,m^{-1}$ at the base to 700 $kN\,m^{-1}$ at the top.

Given that a reasonable estimate of the fundamental mode of vibration is a parabola, obtain the equivalent mass of the tower, M_e, and the equivalent stiffness, K_e (recall Example 1.3). Hence, find the natural period. The shield is assumed to provide all the structural strength as a firmly founded cantilever.

The Strouhal number relationship for a circular cylinder is

$$S = fD/V \approx 0.20$$

where f is the frequency of vortex shedding (in hertz), D is the outer diameter of the cylinder and V is the wind velocity. Ignoring the fact that the wind velocity will vary up the tower, take values of $V = 10, 20, 30$ and $40\ m\,s^{-1}$, calculate the forcing frequency that this implies, and hence obtain the dynamic amplification factors for the transverse loading due to vortex shedding. The damping level is to be taken as $\zeta = 0.03$.

3.7 The dynamic amplification factors obtained in the previous problem are largest when the wind speed is in the vicinity of $30\ m\,s^{-1}$. In fact resonance occurs when $V = 28\ m\,s^{-1}$ because then the frequency of vortex shedding is identical to the fundamental natural frequency of

111

the shield. The calculations in the previous problem lead to values of the equivalent stiffness and equivalent mass of $K_e = 13\,400$ kN m^{-1} and $M_e = 3400$ kN s^2 m^{-1}, respectively. This leads to a natural frequency of $\omega = 2.0$ rad s^{-1}.

The lift force (that is, the horizontal force perpendicular to the wind direction) on a unit length of tower is given by the expression

$$P_{max} = 0.5(\rho V^2/2)D$$

where a lift coefficient of 0.5 has been used, and is broadly typical of such situations, $\rho = 1.2 \times 10^{-3}$ kN s^2 m^{-4} is the density of air, V is the wind speed and, as before, D is the outer diameter of the shield. The periodicity of the force is identical to that of the vortex shedding.

Find the equivalent lateral load on the shield (i.e. P_e) given that the wind is assumed uniform at 28 m s^{-1}. Hence show the peak lateral dynamic displacement of the shield is about 0.34 m for a damping level of 0.03.

3.8 A temporary riser is planned for a flare boom support tower alongside an oil production platform. The riser, a steel pipe of outside diameter 180 mm and wall thickness 10 mm, is to provide water for a drilling operation and is approximately 150 m long. At its upper end, approximately 25 m above mean water level, it is attached to a constant tension device. This tension, which drops off slightly due to riser self-weight as one goes down it, is countered at the mudline by a heavy collar. It is this tension that supplies almost all the resistance to lateral forces. The riser is effectively a huge taut string, pinned at each end, because the influence of flexural rigidity, given a reasonable tension, is generally insignificant. In the present case the tension supplied is about 100 kN.

The fundamental mode of a taut string has the form,

$$\Delta(z, t) = \Delta_{dyn}(t) \sin(\pi z/l)$$

where Δ_{dyn} is the displacement at midheight and $\sin(\pi z/l)$ is the shape function. The strain energy stored in the deflected string is

$$\tfrac{1}{2}H \int_0^l \left(\frac{\partial \Delta}{\partial z}\right)^2 dz$$

and can be thought of as the work done by the tension, H, in countering the 'axial shortening'

$$\tfrac{1}{2} \int_0^l \left(\frac{\partial \Delta}{\partial z}\right)^2 dz$$

that arises when the string is deflected from the vertical.

112

(a)   Use this information to show that the static midspan deflection of the riser under a uniform current of $1 \, \mathrm{m \, s^{-1}}$ is approximately 2.3 m. Assume a drag coefficient of 0.7.

(b)   Show that the formula for the fundamental period of the riser is

$$T = 2 \sqrt{\left(\frac{ml^2}{H} \right)}$$

where m is the total mass per unit length, consisting of the mass of the pipe, the contained water and the added mass of the seawater based on an added mass coefficient of unity. Show that the period is about 10 s.

In point of fact the riser is far too flexible and a system of stays is necessary. Stays at two levels, dividing the riser into three roughly equal parts, will reduce the current drag deflection and also shorten the period significantly. The stays can be anchored to the flare support structure and the only mode of vibration then possible is the third mode, which has nodes at the ends and the two equally spaced intermediate stay levels. The mode of vibration is

$$\Delta(z, t) = \Delta_{\mathrm{dyn}}(t) \sin(3\pi z / l)$$

and the period is reduced to about 3 s, which puts the riser outside the region of significant excitation from wave action.

4

Duhamel's integral and the numerical solution of the τ oscillator equation

4.1 Introduction

In the first part of this chapter, we consider the exact solution, in the form of an integral, of the equation of motion of a single-degree-of-freedom oscillator responding linearly to a general forcing function. This solution is known as Duhamel's integral. A couple of examples are worked for the express purpose of recovering solutions obtained by other means in earlier chapters.

Duhamel's integral then forms the basis for obtaining an exact recursive relationship to facilitate solutions when the forcing function is sufficiently complicated to exclude the possibility of an analytical solution but is amenable to division into simple piecewise linear segments. However, we do not approach this numerical work head on; rather, we develop it by first looking at very simple differential equations and build on that. The end result is a straightforward, accurate method, which can be extended to cover problems where system non-linearity must be accounted for.

The examples chosen illustrate the general efficacy of the particular method advanced here, which is one of any number of approaches possible. More information on other methods can be obtained in the books by Biggs (1964), Clough and Penzien (1975), and Bathe and Wilson (1976), for example.

A brief discussion of inelastic dynamic analysis is given. This topic probably deserves more emphasis than that implied by the somewhat cursory glance given it at the end of the chapter: nevertheless the example serves a useful purpose in showing that inelastic dynamic analysis is tricky, and time consuming. The practitioner is well advised to weigh the costs and benefits of it carefully.

Finally, while it is true that for the important business of number crunching numerical techniques are very often indispensable, the pre-processing and post-processing are, if anything, more important. In this context

114

pre-processing refers to setting up the equation of motion and deciding on a suitable form for the input forcing function: sufficient has been said of that in earlier chapters for there to be no need to say more. Post-processing refers to the way in which the desired response information is selected and displayed – in short, sensible data management techniques are essential.

4.2 Duhamel's integral

We establish here the general solution, in the form of an integral, to the equation of motion:

$$\ddot{\Delta} + 2\zeta\omega\dot{\Delta} + \omega^2 \Delta = F(t)/M = a(t) \tag{4.1}$$

where $F(t)$ is any explicit function of time, so $a(t)$ is an imposed acceleration. The solution is sought for the case of zero initial conditions.

Probably the most direct path to the solution is that due originally to Lord Rayleigh (1945), and we shall use it. In Chapter 2 the free vibrational response to an initial velocity $\dot{\Delta}_0$ was given as

$$\Delta(t) = (\dot{\Delta}_0/\omega_d)e^{-\zeta\omega t} \sin \omega_d t$$

where $\omega_d = \sqrt{(1 - \zeta^2)}\omega$. Therefore, it follows that the effect at time t of a velocity $V(\tau)$ imparted earlier at time τ is

$$\Delta(t) = (V(\tau)/\omega_d)e^{-\zeta\omega(t-\tau)} \sin [\omega_d(t - \tau)] \qquad t \geqslant \tau \tag{4.2}$$

Now, the effect of an imposed acceleration $a(\tau)$ is to generate in time $d\tau$ an imposed velocity $a(\tau)\, d\tau$, and the result of this at a subsequent time t is

$$d\Delta(t) = (a(\tau)\, d\tau/\omega_d)e^{-\zeta\omega(t-\tau)} \sin [\omega_d(t - \tau)] \tag{4.3}$$

Hence Equation 4.3 allows the solution of Equation 4.1 to be written as

$$\Delta(t) = \int_0^t d\Delta(t) = (1/\omega_d) \int_0^t a(\tau)e^{-\zeta\omega(t-\tau)} \sin [\omega_d(t - \tau)\, d\tau \tag{4.4}$$

The lower limit of the integral is taken as zero for then we automatically meet the stipulated requirement of initial conditions of the form $\Delta(0) = \dot{\Delta}(0) = 0$.

Equation 4.4 is Duhamel's integral. If initial conditions are prescribed the general solution consists of Equation 4.4 plus Equation 2.11.

A particular feature of Duhamel's integral is that we are able formally to separate transients due to the imposition of initial conditions from those due to the sudden application of the forcing function. This clear-cut distinction is generally not possible with other methods of solving the oscillator equation.

Only a relatively few forcing functions are integrable analytically so Duhamel's integral, in the above form, has obvious shortcomings. Nevertheless, there are ways around that problem, as will be discussed after two illustrative analytical cases are presented.

4.2.1 *Response to a suddenly applied load of constant magnitude*

In this case $a(\tau) = F/M$ = constant, for all time. In the special case of zero damping, $\zeta = 0$, the solution is

$$\Delta(t) = (F/M\omega) \int_0^t \sin\left[\omega(t - \tau)\right] \, d\tau \tag{4.5}$$

Let $t' = t - \tau$, so that $d\tau = -dt'$, and we have

$$\Delta(t) = -(F/M\omega) \int_t^0 \sin \omega t' \, dt' = (F/M\omega) \int_0^t \sin \omega t' \, dt'$$

That is

$$\Delta(t) = (F/K)(1 - \cos \omega t) \tag{4.6}$$

which is the solution previously obtained by two quite different methods in Chapters 1 and 2.

The inclusion of damping requires evaluation of the integral

$$\Delta(t) = (F/M\omega_d) \int_0^t e^{-\zeta\omega t'} \sin \omega_d t' \, dt' \tag{4.7}$$

From integral tables

$$\int e^{Az} \sin Bz \, dz = \frac{e^{Az}}{A^2 + B^2} (A \sin Bz - B \cos Bz)$$

So our solution is

$$\Delta(t) = \frac{F}{M\omega_d} \left(\frac{e^{-\zeta\omega t'}}{\zeta^2\omega^2 + \omega_d^2} (-\zeta\omega \sin \omega_d t' - \omega_d \cos \omega_d t') \right) \Bigg|_0^t \tag{4.8}$$

After rearrangement we obtain

$$\Delta(t) = \frac{F}{K} \left[1 - e^{-\zeta\omega t} \left(\cos \omega_d t + \frac{\zeta}{\sqrt{(1 - \zeta^2)}} \sin \omega_d t \right) \right] \tag{4.9}$$

which is the same as Equation 2.15.

116

4.2.2 Response of an undamped oscillator to an exponentially decaying force

Consider an undamped oscillator starting from rest and subject to a force

$$F(t) = F_0 e^{-\alpha t} \tag{4.10}$$

Note first that when α is small the forcing function approaches a constant loading, whereas when α is large the force approaches a spike loading applied at time zero. Figure 4.1 illustrates these trends. Note also that an impulse quantity could be defined as

$$I = \int_0^\infty F_0 e^{-\alpha t}\, dt = F_0/\alpha \tag{4.11}$$

Duhamel's integral is

$$\Delta(t) = (F_0/M\omega) \int_0^t e^{-\alpha\tau} \sin\left[\omega(t-\tau)\right]\, d\tau \tag{4.12}$$

and by again making the substitution $t' = t - \tau$ we arrive at

$$\Delta(t) = \frac{F_0 e^{-\alpha\tau}}{M\omega} \int_0^t e^{\alpha t'} \sin \omega t'\, dt' \tag{4.13}$$

The integral is identical in form to that of Equation 4.7 and so the solution is

$$\Delta(t) = \frac{F_0}{M\omega(\alpha^2 + \omega^2)} (\alpha \sin \omega t - \omega \cos \omega t + \omega e^{-\alpha t}) \tag{4.14}$$

When α becomes small, more precisely when $\alpha \ll \omega$, we have

$$\Delta(t) \rightarrow (F_0/K)(1 - \cos \omega t)$$

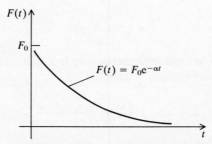

Figure 4.1 Forcing function exhibiting exponential decay.

117

which is the response to a suddenly applied load of constant magnitude. At the other extreme, when $\alpha \gg \omega$, we take the limit letting $I \ (= F_0/\alpha)$ stay constant. This yields

$$\Delta(t) \to \frac{F_0}{M\omega\alpha^2} (\alpha \sin \omega t) \to \frac{I}{M\omega} \sin \omega t$$

which is the response to an impulse, given originally by Equation 2.21.

Clearly, many more examples could be given. Obviously, a good table of integrals is indispensable. The table by Klerer and Grossman (1971) has some 2000 entries, all of which have been checked by computer.

4.3 The numerical solution of equations of motion

4.3.1 Numerical solution of the equation $\ddot{\Delta}(t) = a(t)$

The interest here first centres on the development of an approximate recurrence relationship for the numerical solution of the equation

$$\ddot{\Delta}(t) = a(t) \tag{4.15}$$

subject to initial conditions of $\Delta(0) = \dot{\Delta}(0) = 0$. The above equation is just a part of the full oscillator eqation – damping force and spring force terms having been deliberately omitted from the left-hand side. The right-hand side represents forcing in the form of a known acceleration.

Suppose we are interested in obtaining the solution at discrete times t_i, t_j, t_k, . . ., spaced a time τ apart, that is, $t_j - t_i = \tau$, $t_k - t_j = \tau$, etc. Suppose, further, that at time t_j the forcing function is known, namely $a(t_j) = a_j$. At time t_j the response velocity is assumed to be the average over the adjacent time intervals, namely,

$$\dot{\Delta}_j = \dot{\Delta}(t_j) \approx \left(\frac{\Delta_k - \Delta_j}{\tau} + \frac{\Delta_j - \Delta_i}{\tau} \right)/2$$

$$\approx (\Delta_k - \Delta_i)/2\tau$$

where Δ_i, Δ_j and Δ_k are self-explanatory. In this central difference formulation the response acceleration is then

$$\ddot{\Delta}_j = \ddot{\Delta}(t_j) \approx \left(\frac{\Delta_k - \Delta_j}{\tau} - \frac{\Delta_j - \Delta_i}{\tau} \right)/\tau$$

$$\approx (\Delta_k - 2\Delta_j + \Delta_i)/\tau^2$$

This leads to an approximation recursion formula of the form

$$(\Delta_k - 2\,\Delta_j + \Delta_i)/\tau^2 \approx a_j$$

or

$$\Delta_k \approx a_j\tau^2 + 2\,\Delta_j - \Delta_i$$

which holds for $k = 1, 2, 3, 4, \ldots$, etc.

When $k = 2$, we need to know Δ_1 and Δ_0 in order to find Δ_2. The value of Δ_0 is zero, as given by our initial condition. When $k = 1$, we have

$$\Delta_1 \approx a_0\tau^2 + 2\,\Delta_0 - \Delta_{-1}$$
$$\approx a_0\tau^2 - \Delta_{-1}$$

From the initial condition on velocity, namely, $\dot\Delta(0) = 0$, we have

$$\Delta_1 = \Delta_{-1}$$

and so,

$$\Delta_1 \approx a_0\tau^2/2$$

Hence, we have as the required approximate solution to Equation 4.15, the self-starting sequence

$$\Delta_0 = 0$$
$$\Delta_1 = a_0\tau^2/2$$
$$\Delta_k = a_j\tau^2 + 2\,\Delta_j - \Delta_i \qquad k = 2, 3, 4, \ldots \qquad (4.16)$$

This solution increases in accuracy as τ decreases.

Nevertheless, in the special case where the forcing function is piecewise linear, an exact solution to Equation 4.15 may be found. Since this is useful later we now spend some time elucidating that solution.

Piecewise linear means that the forcing function is straight between time stations, as in Figure 4.2. If the time interval τ is sufficiently small, almost any actual variation in forcing can be accurately represented by a sequence of straight lines.

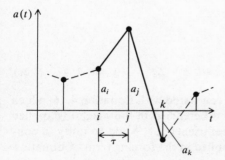

Figure 4.2 Definition diagram for piecewise linear forcing.

In the first time interval, $0 \leqslant t \leqslant \tau$, the equation is

$$\ddot{\Delta} = a_0 + (a_1 - a_0)t/\tau \qquad (4.17)$$

subject to $\Delta(0) = \dot{\Delta}(0) = 0$. The solution is

$$\dot{\Delta}(t) = a_0 t + \tfrac{1}{2}(a_1 - a_0)t^2/\tau$$

and

$$\Delta(t) = \tfrac{1}{2} a_0 t^2 + \tfrac{1}{6}(a_1 - a_0)t^3/\tau$$

yielding at $t = \tau$

$$\dot{\Delta}_1 = \tfrac{1}{2} a_0 \tau + \tfrac{1}{2} a_1 \tau$$

and

$$\Delta_1 = \tfrac{1}{3} a_0 \tau^2 + \tfrac{1}{6} a_1 \tau^2 \qquad (4.18)$$

In the second time interval we have to solve

$$\ddot{\Delta} = a_1 + (a_2 - a_1)t/\tau$$

subject to $\Delta(0) = \Delta_1$, $\dot{\Delta}(0) = \dot{\Delta}_1$. We obtain

$$\Delta_2 = (\tfrac{1}{6} a_0 + \tfrac{2}{3} a_1 + \tfrac{1}{6} a_2)\tau^2 + 2\,\Delta_1 - \Delta_0 \qquad (4.19)$$

and, even though $\Delta_0 = 0$, we have written the solution as in Equation 4.19 to emphasise a similarity with Equation 4.16.

Proceeding to the third stage we find that

$$\Delta_3 = (\tfrac{1}{6} a_1 + \tfrac{2}{3} a_2 + \tfrac{1}{6} a_3)\tau^2 + 2\,\Delta_2 - \Delta_1$$

By a process of induction, then, we are led to the general, exact, solution of Equation 4.15 for piecewise linear forcing, which is

$$\Delta_0 = 0$$

$$\Delta_1 = (\tfrac{1}{3} a_0 + \tfrac{1}{6} a_1)\tau^2$$

$$\Delta_k = (\tfrac{1}{6} a_i + \tfrac{2}{3} a_j + \tfrac{1}{6} a_k)\tau^2 + 2\,\Delta_j - \Delta_i \qquad k = 2, 3, 4, .. \quad (4.20)$$

Equation 4.20, which is exact, is to be compared with Equation 4.16, which is approximate. One sees that the only difference is in the weighting applied to the forcing terms: the numerical coefficients $\tfrac{1}{6}$, $\tfrac{2}{3}$, $\tfrac{1}{6}$ add to unity in conformity with the coefficient of unity implied in the forcing term of Equation

4.16. Furthermore, the central difference form for response acceleration, $(\Delta_k - 2\,\Delta_j + \Delta_i)/\tau^2$, has been preserved in the exact solution, and this has a simplifying effect on calculations.

In fact, what is implied by the form of Equation 4.20 is that significant improvements in solution accuracy may be gained quite simply by operating on the right-hand side of the equation of motion, rather than concentrating on the left-hand side of the equation of motion. This portion of the book, therefore, differs from traditional introductory treatments of numerical routines for the oscillator equation, where the reverse procedure is common and, because of this, is often less straightforward.

4.3.2 Numerical solution of the equation $\ddot{\Delta} + \omega^2\,\Delta = a(t)$

The undamped oscillator equation

$$\ddot{\Delta} + \omega^2\,\Delta = a(t) \tag{4.21}$$

is to be solved, for the case of zero initial conditions, using an approximate recursive relationship based on the exact one from Equation 4.20, the implication being that the solution routine is a good one – a point that will be confirmed later in the next subsection.

At time t_j the left-hand side of Equation 4.21 is given approximately by

$$\frac{(\Delta_k - 2\,\Delta_j + \Delta_i)}{\tau^2} + \omega^2\,\Delta_j$$

whereas we choose to write the right-hand side as

$$(\tfrac{1}{6}a_i + \tfrac{2}{3}a_j + \tfrac{1}{6}a_k)$$

This leads to a solution sequence (for zero initial conditions) of

$$\Delta_0 = 0$$

$$\Delta_1 \approx (\tfrac{1}{3}a_0 + \tfrac{1}{6}a_1)\tau^2$$

$$\Delta_k \approx (\tfrac{1}{6}a_i + \tfrac{2}{3}a_j + \tfrac{1}{6}a_k)\tau^2 + [2 - (\omega\tau)^2]\,\Delta_j - \Delta_i \qquad k = 2, 3, 4, \ldots \tag{4.22}$$

This is an approximate numerical routine for the undamped oscillator equation responding to piecewise linear forcing. It will be an accurate representation of the solution if τ is sufficiently small for the forcing function to be closely approximated by piecewise linear segments and if, in addition, τ is a small fraction of the natural period T. This latter is achieved, in general, if $\omega\tau < 1$: a typical upper limit on τ is $\tau/T \approx 0.1$.

121

Example 4.1 Response to a sinusoidal pulse

The sinusoidal pulse shown in Figure 4.3 is of the form

$$a(t) = \begin{cases} 1 \sin \Omega t & 0 \leqslant t \leqslant 1.2 \text{ s} \\ 0 & t > 1.2 \text{ s} \end{cases}$$

where $\Omega = 2\pi/2.4 = 2.618 \text{ rad s}^{-1}$. Let us suppose that the natural frequency of the oscillator is also 2.618 rad s^{-1}. That is, $\omega = 2.618 \text{ rad s}^{-1}$ or the natural period is $T = 2.4$ s.

We are interested in the peak displacement of the oscillator in responding to this loading. Accordingly, we ignore damping because the peak will occur early in the response.

EXACT SOLUTION

Before calculating the response numerically we first establish the exact solution (via Duhamel's integral, for example). The undamped equation of motion for the case when $\Omega \equiv \omega$ is

$$\ddot{\Delta} + \omega^2 \Delta = \sin \omega t \qquad 0 \leqslant t \leqslant \pi/\omega$$

subject to $\Delta(0) = \dot{\Delta}(0) = 0$. The solution is

$$\Delta(t) = \frac{1}{2\omega} \left(\frac{1}{\omega} \sin \omega t - t \cos \omega t \right) \qquad t \leqslant \frac{\pi}{\omega}$$

Notice that the response velocity is

$$\dot{\Delta}(t) = \frac{1 \times t \times \sin \omega t}{2} \qquad t \leqslant \frac{\pi}{\omega}$$

and that when $t = \pi/\omega$ (i.e. $t = 1.2$ s) the velocity is zero. Thus the peak displacement response occurs right at the end of the pulse, which is rather convenient because we do not have to consider the free vibration after the pulse in order to ascertain the maximum.

Since the period is 2.4 s a time increment of, say, $\tau = 0.2$ s is more than satisfactory because $\tau/T = 1/12$. The exact solution for displacement at each time increment is given in Table 4.1. These results will now be compared with the approximate values obtained in Equation 4.22.

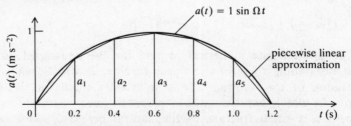

Figure 4.3 A sinusoidal pulse and an equivalent piecewise linear representation.

122

Table 4.1 Response to a sinusoidal pulse.

t(s)	Δ(m)
0	$\Delta_0 = 0$
0.2	$\Delta_1 = 0.0034$
0.4	$\Delta_2 = 0.0250$
0.6	$\Delta_3 = 0.0729$
0.8	$\Delta_4 = 0.1396$
1.0	$\Delta_5 = 0.2019$
1.2	$\Delta_6 = 0.2292 = \Delta_{max}$

APPROXIMATE NUMERICAL SOLUTION

The input acceleration is $1 \sin \Omega t$ and at the discrete time intervals chosen we have

$$a_0 = 0 \qquad\qquad a_4 = 0.8660 \text{ m s}^{-2}$$
$$a_1 = 0.5 \text{ m s}^{-2} \qquad a_5 = 0.5 \text{ m s}^{-2}$$
$$a_2 = 0.8660 \text{ m s}^{-2} \quad a_6 = 0$$
$$a_3 = 1 \text{ m s}^{-2} \qquad a_7 = 0, \text{ etc.}$$

The recursive relationship is

$$\Delta_0 = 0$$

$$\Delta_1 = (\tfrac{1}{3}a_0 + \tfrac{1}{6}a_1) \times 0.04$$

$$\Delta_k = (\tfrac{1}{6}a_i + \tfrac{2}{3}a_j + \tfrac{1}{6}a_k) \times 0.04 + 1.7258\,\Delta_j - \Delta_i \qquad \text{for } k = 2, 3, 4, \dots$$

Working through this gives

$$\Delta_0 = 0$$

$$\Delta_1 = (0 + \tfrac{1}{6} \times 0.5) \times 0.04 = 0.0033 \text{ m}$$

$$\Delta_2 = 0 + \tfrac{2}{3} \times 0.5 + \tfrac{1}{6} \times 0.8660) \times 0.04 + 1.7258 \times 0.0033 - 0$$
$$= 0.0248 \text{ m}$$

$$\Delta_3 = (\tfrac{1}{6} \times 0.5 + \tfrac{2}{3} \times 0.8660 + \tfrac{1}{6} \times 1) \times 0.04 + 1.7258 \times 0.0248 - 0.0033$$
$$= 0.0726 \text{ m}$$

By which continuing process we obtain

$$\Delta_4 = 0.1387 \text{ m} \qquad \Delta_6 = 0.2254 \text{ m}$$

$$\Delta_5 = 0.1999 \text{ m} \qquad \Delta_7 = 0.1924 \text{ m, etc}$$

When hand calculations are being done it is important to check calculations regularly since errors propagate through the solution.

The peak displacement occurs at 1.2 s and is 0.2254 m, which compares very favourably with the exact value of 0.2292 m. The numerical answers are all slightly less than their analytical counterparts and this, in fact, is mainly due to the input acceleration for the numerical scheme being slightly less than the input for the analytical scheme everywhere, except at the chosen time intervals. There is obviously not much in it though.

The above numerical calculation could proceed indefinitely past Δ_6 (note that we obtained Δ_7 to prove that Δ_6 was the largest): it is simply a free vibration of amplitude 0.2254 m. In the case of the earlier analytical solution, a different solution – that for free vibration under an 'initial' displacement – would have to be found.

Example 4.2   Response to a form of impulsive loading

The loading shown in Figure 4.4 is fed through the same oscillator as in Example 4.1. We again ignore damping and take a time increment of $\tau = 0.2$ s.

The response is

$$\Delta_0 = 0$$

$$\Delta_1 = 0.04(\tfrac{1}{3} \times 0 + \tfrac{1}{6} \times 1) = 0.0067 \text{ m}$$

$$\Delta_2 = 0.04(\tfrac{1}{6} \times 0 + \tfrac{2}{3} \times 1 + \tfrac{1}{6} \times -2) + 1.7258 \times 0.0067 - 0 = 0.0249 \text{ m}$$

$$\Delta_3 = 0.04(\tfrac{1}{6} \times 1 + \tfrac{2}{3} \times -2 + \tfrac{1}{6} \times 0.5) + 1.7258 \times 0.0249 - 0.0067$$
$$= -0.0070 \text{ m}$$

and so on. The result is shown plotted in Figure 4.5. The peak displacement is 0.2095 m occurring at a time 1.8 s after the start. The response bears little resemblance, interval-by-interval, to the loading for the principal reason that the time of loading is relatively short *vis-à-vis* the period of the system.

A closed-form analytical solution for the above problem is quite unwieldy (although, as we shall shortly see, it is possible to obtain an exact solution as a recursion formula via Duhamel's integral). Consequently we shall exploit the fact

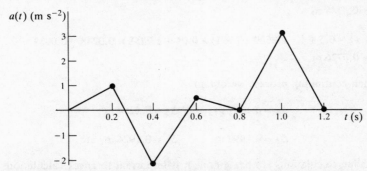

Figure 4.4 An 'impulsive' loading in the form of a series of spikes.

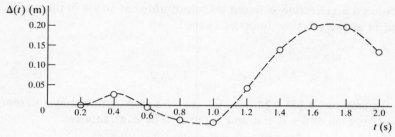

Figure 4.5 Response to spike loading.

that in a physical sense the loading resembles an impulse as a means of providing a check on the numerical solution. The impulse of the loading is (via the trapezoidal rule):

$$I = M(\tfrac{1}{2}a_0 + a_1 + a_2 + a_3 + a_4 + a_5 + \tfrac{1}{2}a_6)\tau$$

where M is the mass of the oscillator. This yields

$$I = 0.5M \text{ kN s}$$

From Equation 2.22 we have a peak displacement of

$$\Delta_{max} = \frac{I}{M\omega} = \frac{0.5}{2.618} = 0.191 \text{ m}$$

an answer that lends support to the view that the loading is largely impulsive when applied to a system with a period of 2.4 s. This 'impulsive' loading takes 1.2 s to be applied, after which a time of $T/4 = 0.6$ s is required for the response to reach its maximum. The total elapsed time is $1.2 + 0.6 = 1.8$ s, which also compares favourably with the facts.

4.3.3 An exact recursion formula for the equation $\ddot{\Delta} + 2\zeta\omega\dot{\Delta} + \omega^2 \Delta = a(t)$

Using Duhamel's integral, Dempsey and Irvine (1978) have shown that an exact recursion relation for the response of a single-degree-of-freedom oscillator, starting with zero initial conditions, is given by

$$\Delta_0 = 0$$

$$\Delta_1 = Ca_0 + Da_1$$

$$\Delta_k = [Aa_i + (C - B)a_j + Da_k] + 2e^{-\zeta\omega\tau} \cos(\omega_d\tau) \Delta_j - e^{-2\zeta\omega\tau} \Delta_i$$

$$k = 2, 3, 4, \ldots \qquad (4.23)$$

for displacement response. The result for vibrational velocity is

$$\dot{\Delta}_0 = 0$$

$$\dot{\Delta}_j = \omega_d e^{\zeta\omega\tau} \operatorname{cosec}(\omega_d\tau)\{-Ca_j - Da_k - [1 - \omega^2(C + D)] \Delta_j + \Delta_k\}$$

$$j = 1, 2, 3, \ldots \qquad (4.24)$$

125

Vibrational acceleration is found by substituting the above in the oscillator equation at the jth time interval. It reads

$$\ddot{\Delta}_0 = a_0$$
$$\ddot{\Delta}_j = a_j - 2\zeta\omega\dot{\Delta}_j - \omega^2\Delta_j \qquad j = 1, 2, 3, \ldots \tag{4.25}$$

For Equations 4.23–4.25, ζ is the fraction of critical damping $\omega_d = \sqrt{(1 - \zeta^2)}\omega$ and τ is the time interval. The constants are

$$A = \frac{1}{\omega^2} e^{-\zeta\omega\tau} \left[\left(1 + \frac{2\zeta}{\omega\tau}\right) e^{-\zeta\omega\tau} - \left(\frac{2\zeta}{\omega\tau}\cos \omega_d\tau + \frac{1 - 2\zeta^2}{\omega_d\tau}\sin \omega_d\tau\right)\right]$$

$$B = A + \frac{1}{\omega^2} e^{-\zeta\omega\tau} \left(\cos \omega_d\tau - \frac{\zeta}{\sqrt{(1 - \zeta^2)}}\sin \omega_d\tau - e^{-\zeta\omega\tau}\right)$$

$$\text{(4.26)}$$

$$C = \frac{2\zeta}{\omega^3\tau} - \frac{1}{\omega^2} e^{-\zeta\omega\tau} \left[\left(1 + \frac{2\zeta}{\omega\tau}\right)\cos \omega_d\tau - \left(\frac{1 - 2\zeta^2}{\omega_d\tau} - \frac{\zeta}{\sqrt{(1 - \zeta^2)}}\right)\sin \omega_d\tau\right]$$

$$D = -C + \frac{1}{\omega^2} \left[1 - e^{-\zeta\omega\tau}\left(\cos \omega_d\tau + \frac{\zeta}{\sqrt{(1 - \zeta^2)}}\sin \omega_d\tau\right)\right]$$

and need to be evaluated once only for any particular problem.

The solutions as outlined above are general and hold for any time step, provided, of course, that the forcing is linear between time stations. Damping is included and, although the constants A to D are somewhat involved, the only effort required is to code them up for machine usage.

If there are initial conditions, as well, we simply add the exact solutions for those, as given by Equation 2.11, to the above results.

It is a significant point that the above method is exact for any time step, provided the input is still thereby accurately prescribed. In a moment or two we shall extract various special cases from the above results as a means of tying in the earlier work of this section, and we shall propose an approximate solution for the oscillator equation for use when hand methods are appropriate. In this latter case it is necessary to place limits on the time step, and we do this to ensure reasonable accuracy. As the time increment becomes smaller, the answer should converge to the exact solution. If the time step is small then, at the same time, errors obtaining at one station will not adversely affect the accuracy of the calculation at the next time step by, for example, becoming much larger: that is, the numerical scheme will be stable. Matters of convergence and stability are central to all approximate numerical schemes, but this is not the appropriate place for further discussion. Enough has been said: in most cases where a problem might exist, a remedy is to reduce the time step.

EXACT SOLUTION FOR ZERO DAMPING

Setting $\zeta = 0$ in Equations 4.23 and 4.26 yields the exact solution for the oscillator equation

$$\ddot{\Delta} + \omega^2 \Delta = a(t)$$

with $\Delta_0 = \dot{\Delta}_0 = 0$, as

$$\Delta_0 = 0$$

$$\Delta_1 = \frac{1}{\omega^2} \left(\frac{\sin \omega\tau}{\omega\tau} - \cos \omega\tau \right) a_0 + \frac{1}{\omega^2} \left(1 - \frac{\sin \omega\tau}{\omega\tau} \right) a_1$$

$$\Delta_k = \frac{1}{\omega^2} \left(1 - \frac{\sin \omega\tau}{\omega\tau} \right) a_i + \frac{2}{\omega^2} \left(\frac{\sin \omega\tau}{\omega\tau} - \cos \omega\tau \right) a_j + \frac{1}{\omega^2} \left(1 - \frac{\sin \omega\tau}{\omega\tau} \right) a_k$$
$$+ 2 \cos \omega\tau \, \Delta_j - \Delta_i \qquad k = 2, 3, 4, \ldots \qquad (4.27)$$

These equations are easily used to give exact results for Example 4.2 — the case of semi-impulsive spike loading. The exact results are

$$\Delta_0 = 0 \qquad\qquad \Delta_6 = 0.0573 \text{ m}$$

$$\Delta_1 = 0.0066 \text{ m} \qquad\qquad \Delta_7 = 0.1516 \text{ m}$$

$$\Delta_2 = 0.02418 \text{ m} \qquad\qquad \Delta_8 = 0.2053 \text{ m}$$

$$\Delta_3 = -0.0067 \text{ m} \qquad\qquad \Delta_9 = 0.2039 \text{ m}$$

$$\Delta_4 = -0.0360 \text{ m} \qquad\qquad \Delta_{10} = 0.1478 \text{ m}$$

$$\Delta_5 = -0.0326 \text{ m} \qquad\qquad \text{etc.}$$

By comparison, the approximate answers of Example 4.2 agree to within 2%; this is better accuracy than that to which the input will be known in any practical problem. This point should be constantly borne in mind in all calculations.

APPROXIMATE SOLUTION FOR ZERO DAMPING AND A SMALL TIME STEP

When, in addition to ζ being zero, we stipulate that $\omega\tau \ll 1$, we may effect further simplifications in Equations 4.26. Noting that when $\omega\tau \ll 1$

$$\sin \omega\tau \approx \omega\tau - \tfrac{1}{6}(\omega\tau)^3, \qquad \cos \omega\tau \approx 1 - \tfrac{1}{2}(\omega\tau)^2$$

we obtain

$$\Delta_0 = 0$$

$$\Delta_1 \approx (\tfrac{1}{3}a_0 + \tfrac{1}{6}a_1)\tau^2$$

$$\Delta_k \approx (\tfrac{1}{6}a_i + \tfrac{2}{3}a_j + \tfrac{1}{6}a_k)\tau^2 + [2 - (\omega\tau)^2]\,\Delta_j - \Delta_i \qquad k = 2, 3, 4, \ldots \qquad (4.28)$$

These are, of course, identical to the solutions given by Equation 4.22 and we have come full circle in that regard. A more accurate rendition would include the next terms in the series expansions for the sine and cosine functions: such terms contain further products and powers of ω and τ.

The above solution is accurate when $\omega\tau$ is small: it becomes exact when $\omega = 0$, which is the case first treated in this numerical work (recall Sec. 4.3.1), and again we have come full circle.

APPROXIMATE SOLUTION WITH DAMPING AND A SMALL TIME STEP

As an adjunct to Equations 4.23–4.26, it is useful to have an approximate solution, readily amenable to hand calculation, which includes damping and is therefore of wider applicability than Equation 4.22, which does not. The equation of motion is

$$\ddot{\Delta} + 2\zeta\omega\dot{\Delta} + \omega^2\Delta = a(t)$$

and we note that the left-hand side may be written

$$\ddot{\Delta}_j \approx (\Delta_k - 2\Delta_j + \Delta_i)/\tau^2$$

$$2\zeta\omega\dot{\Delta}_j \approx \zeta\omega(\Delta_k - \Delta_i)/\tau$$

and

$$\omega^2\Delta_j \equiv \omega^2\Delta_j$$

This leads to the solution

$$\Delta_0 = 0$$

$$\Delta_1 = (\tfrac{1}{3}a_0 + \tfrac{1}{6}a_1)\tau^2$$

$$(1 + \zeta\omega\tau)\,\Delta_k = (\tfrac{1}{6}a_i + \tfrac{2}{3}a_j + \tfrac{1}{6}a_k)\tau^2 + [2 - (\omega\tau)^2]\,\Delta_j - (1 - \zeta\omega\tau)\,\Delta_i$$
$$k = 2, 3, 4, \ldots \qquad (4.29)$$

This solution, which could be called the weighted force method or, perhaps, *linear* force method, will generally give good answers when

$\omega\tau \ll 1$. It is only slightly more involved than the traditional approximate solution, called by Biggs (1964) the constant velocity method, given by

$$\Delta_0 = 0$$

$$\Delta_1 = \tfrac{1}{2}a_0\tau^2$$

$$(1 + \zeta\omega\tau)\,\Delta_k = a_j\tau^2 + [2 - (\omega\tau)^2]\Delta_j - (1 - \zeta\omega\tau)\Delta_i \qquad k = 2, 3, 4, \ldots \quad (4.30)$$

It is, however, more accurate than that method because of the weighting applied to the right-hand side, and in this respect rivals the so-called linear acceleration method, in which the *response* acceleration, rather than the imposed acceleration, is assumed to be linear within a time step. (See Biggs (1964) and Clough and Penzien (1975) for further discussion of the linear acceleration method.) A computer program listing, which incorporates the recursive relationship, Equation 4.29, is given in Section 4.5.

A final point, before we work a couple of examples, is to note that there are other weighting schemes possible. One could, for example, assume that the forcing function was parabolic between time intervals. An exact solution based on this premise would be forthcoming if the method of Dempsey and Irvine (1978) was followed. However, this seems to be splitting hairs somewhat as the extra accuracy is probably more apparent than real, while the extra effort involved in obtaining the solution would be definitely real!

A second and more profitable alternative is to construct a solution for forcing as shown in Figure 4.6a. Labelling levels as a_0', a_1', etc., it is possible to show (see Problem 4.2) that the exact solution to the equation

$$\ddot{\Delta} = a(t)$$

with $\Delta_0 = \dot{\Delta}_0 = 0$, is

$$\Delta_0 = 0$$

$$\Delta_1 = \tfrac{1}{2}a_0'\tau^2$$

$$\Delta_k = \tfrac{1}{2}(a_i' + a_j')\tau^2 + 2\,\Delta_j - \Delta_i \qquad k = 2, 3, 4, \ldots \quad (4.31)$$

This solution, which holds for forcing in the form of a series of short-lived step loadings, may then be adapted directly to the case of input information supplied at regular intervals (viz. a_0, a_1, a_2, etc.). The diagram in Figure 4.6b shows how. Thus, we may surmise that

$$a_0' \equiv \tfrac{1}{2}(a_0 + a_1)$$

$$a_1' \equiv \tfrac{1}{2}(a_1 + a_2)$$

$$a_2' \equiv \tfrac{1}{2}(a_2 + a_3), \text{ etc.}$$

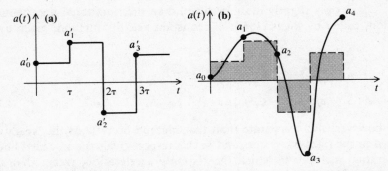

Figure 4.6 Discretisation for the constant average force method: (a) a series of short-lived step loadings; (b) adaptation to input information at regular intervals.

and so another good approximate solution to the damped oscillator responding from rest is

$$\Delta_0 = 0$$

$$\Delta_1 = (\tfrac{1}{4}a_0 + \tfrac{1}{4}a_1)\tau^2$$

$$(1 + \zeta\omega\tau)\,\Delta_k = (\tfrac{1}{4}a_i + \tfrac{1}{2}a_j + \tfrac{1}{4}a_k)\tau^2 + [2 - (\omega\tau)^2]\Delta_j - (1 - \zeta\omega\tau)\Delta_i$$
$$k = 2, 3, 4, \ldots \qquad (4.32)$$

This solution, which might be called the constant average force method, may have advantages over the solution given by Equation 4.29 when the input is not piecewise linear but, rather, is some form of smoother curve that may be approximated by an area-conserving discretisation.

Example 4.3   Earthquake analysis of a 25-storey frame

This example represents a continuation of Example 1.3 wherein a 25-storey K-braced frame was analysed for idealised wind loading. This same frame is now analysed for the effects of one recorded earthquake. The example first serves to show how earthquake loading can be incorporated into the analysis procedure — for the problem is akin to that considered in the previous chapter on response to support motions — it provides a good example of how our numerical schemes can be utilised and, finally, it allows comparison with wind loading in terms of the magnitude of the dynamic response.

Equation 1.20, minus damping, reads

$$\frac{d}{dt}(KE + PE) = P(t)\,\dot{\Delta}$$

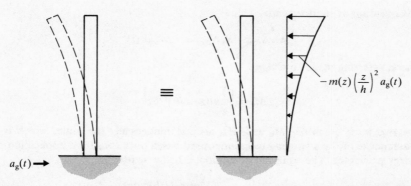

$$-m(z)\left(\frac{z}{h}\right)^2 a_g(t)$$

$$a_g(t) \rightarrow$$

Figure 4.7 Equivalent earthquake loading on the frame: (a) ground acceleration; (b) equivalent distributed loading.

Recalling Example 1.3, we note that the left-hand side of this rate equation is

$$\dot{\Delta}_{\text{dyn}}(M_e \ddot{\Delta}_{\text{dyn}} + K_e \Delta_{\text{dyn}})$$

and so our equation of motion is

$$M_e \ddot{\Delta}_{\text{dyn}} + K_e \Delta_{\text{dyn}} = P(t)$$

where $P(t)$ is the effective total earthquake loading. Figure 4.7 shows the equivalent situation – namely, the ground acceleration, $a_g(t)$, replaced by an equivalent distributed loading.

To arrive at the form of $P(t)$ in this instance, we first form the expression for the increment in work done when the distributed loading

$$-m(z)a_g(t)$$

acts through an increment in dynamic displacement $d\Delta$. Because $\Delta(z,t)$ varies up the tower, we have

$$d\Delta = (z/h)^2 \, d\Delta_{\text{dyn}}$$

where, it will be recalled, $(z/h)^2$ is the shape function for the curve of dynamic deflection. Thus the increment in work done for the whole frame is

$$dW = \left(\int_0^h -m(z)a_g(t)(z/h)^2 \, dz \right) d\Delta_{\text{dyn}}$$

and so, from Equation 1.20, we conclude that

$$P(t) = - \left(\int_0^h m(z)(z/h)^2 \, dz \right) a_g(t)$$

131

The equation of motion reads

$$M_e \ddot{\Delta}_{dyn} + K_e \Delta_{dyn} = -M_{eq} a_g(t)$$

where, referring again to Chapter 1,

$$M_e = \int_0^h m(z)(z/h)^4 \, dz$$

The two mass quantities M_e and M_{eq} are not numerically the same, which is a consequence of the averaging techniques used based on a consistent application of energy principles. The equation of motion is better written as

$$\ddot{\Delta}_{dyn} + \omega^2 \, \Delta_{dyn} = -\left(\frac{\int_0^h m(z)(z/h)^2 \, dz}{\int_0^h m(z)(z/h)^4 \, dz}\right) a_g(t)$$

The dimensionless quantity in brackets on the right is often called the participation factor. In the present problem its value is 5/3, as simple integration indicates.

Thus, if we now include damping, which we are at liberty to do at this stage, the equation of motion reads

$$\ddot{\Delta}_{dyn} + 2\zeta\omega\dot{\Delta}_{dyn} + \omega^2\Delta_{dyn} = -\tfrac{5}{3} a_g(t)$$

where $\omega = \sqrt{(K_e/M_e)} = \sqrt{(1.46 \times 10^3/2.94 \times 10^2)} = 2.23 \text{ rad s}^{-1}$. Damping is chosen as $\zeta = 2\%$. The initial conditions are

$$\Delta_{dyn}(0) = \dot{\Delta}_{dyn}(0) = 0$$

The input, $a_g(t)$, is the well known digitised record of the Taft, California, 1952, earthquake with a peak acceleration of $0.178g$. The record is given as piecewise linear segments spaced at 0.01 s intervals. Choosing this as the integration step is more than satisfactory as the period of the structure is 2.81 s: thus $\omega\tau = 0.0223$.

Because of the number of calculations to be performed a computer is necessary, but Equation 4.23 forms the basis for a simple, exact routine. The output record is shown for the central portion of strong response in Figure 4.8. Because the input is so variable, response is akin to that of a series of 'random' impulses: there is little evidence of periodicity in the forcing function, so response is largely transient and, as is to be expected, at a period close to the natural period of the structure. This is typical of earthquake response.

The peak relative response displacement is $\Delta_{dyn,max} = 0.23$ m, which occurs approximately 15 s after the start. The design wind drift, Δ_{des}, was 0.20 m, so this seismic event is slightly more severe. Notice also from Figure 4.8 that there are several peaks in response near that level.

The peak overturning moment due to the earthquake is related in a simple way to Δ_{des}: it is necessary only to scale the overturning moment due to the design wind loading M_0 ($= 4.25 \times 10^4$ kN m) by the ratio $\Delta_{dyn,max}/\Delta_{des}$, where Δ_{des} is the design wind drift. Thus,

$$M_{eq} = \frac{\Delta_{dyn,max}}{\Delta_{des}} M_0$$

Similarly, the peak seismic base shear is

$$S_{eq} = \frac{\Delta_{dyn,max}}{\Delta_{des}} S_0$$

where S_0 (due to the wind) is 7.5×10^2 kN.

The results are

$$M_{eq} = 4.89 \times 10^4 \text{ kN m}, \qquad S_{r,q} = 8.62 \times 10^2 \text{ kN}$$

The building weight tributary to one frame is 1.44×10^4 kN so, in terms of a fraction of the weight of the building, the peak seismic base shear is 0.06. Compared to the value implied by the peak ground acceleration, viz. 0.178, we see that the building does not respond strongly to the earthquake, which, in this case, has frequencies that are higher than the system frequency. This filtering action is typical of tall buildings and is predicted in a qualitative sense by the response spectrum curves for steady-state response in Figure 3.2.

In a quantitative sense, however, we need to do a little more work. In the steady-state response calculations of Chapter 3, the oscillator responds at the frequency of forcing whereas, as Figure 4.8 has shown, response to the essentially random series of spikes that constitute earthquake ground motion gives rise to response at the structure's natural frequency. We can exploit this fact to help explain the filtering action demonstrated above. For this it may be necessary to refer back to the material on impulsive loading in Chapter 2. The approach now outlined is due to D. E. Hudson.

The undamped displacement response after an impulse I_i delivered at time t_i is

$$\Delta(t) = (I_i/M\omega) \sin \left[\omega(t - t_i) \right] \qquad t \geqslant t_i$$

The force generated is $K \Delta(t)$ or

$$F = K \Delta(t) = \omega I_i \sin \left[\omega(t - t_i) \right]$$

After N such impulses we have, from superposition,

$$F_N = \omega \sum_{i=1}^{N} I_i \sin \left[\omega(t - t_i) \right]$$

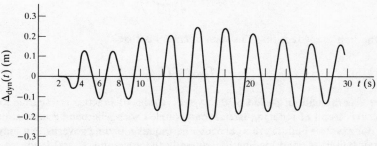

Figure 4.8 Response of a 25-storey frame to the Taft, 1952, earthquake.

or, by the addition formula for trigonometric functions,

$$F_N = \omega \sum_{i=1}^{N} (I_i \sin \omega t \cos \omega t_i - I_i \cos \omega t \sin \omega t_i)$$

$$= \omega \left[\sin (\omega t) \left(\sum_{i=1}^{N} I_i \cos \omega t_i \right) - \cos(\omega t) \left(\sum_{i=1}^{N} I_i \sin \omega t_i \right) \right]$$

This may be rearranged further to

$$F_N = \omega \sqrt{ \left[\left(\sum_{i=1}^{N} I_i \cos \omega t_i \right)^2 + \left(\sum_{i=1}^{N} I_i \sin \omega t_i \right)^2 \right] } \sin (\omega t - \phi)$$

where ϕ is a phase angle given by

$$\tan \phi = \left(\sum_{i=1}^{N} I_i \sin \omega t_i \right) \Big/ \left(\sum_{i=1}^{N} I_i \cos \omega t_i \right)$$

Note that irrespective of ϕ, the peak value of $\sin(\omega t - \phi)$ is unity. Thus the peak force possible after N impulses is

$$F_{\max} = \omega \sqrt{ \left[\left(\sum_{i=1}^{N} I_i \cos \omega t_i \right)^2 + \left(\sum_{i=1}^{N} I_i \sin \omega t_i \right)^2 \right] }$$

That expression may also be written as

$$F_{\max} = \omega \sqrt{ \left(\sum_i I_i^2 + 2 \sum_i \sum_j I_i I_j \cos \left[\omega(t_i - t_j) \right] \right) } \qquad i \neq j$$

and the argument is that the double summation term will be small since $\cos \left[\omega(t_i - t_j) \right]$ will be both positive and negative repeatedly. Consequently, it is reasonable to assume that

$$F_{\max} \approx \omega \sqrt{ \left(\sum_{i=1}^{N} I_i^2 \right) }$$

and the important fact that this then uncovers is that

$$F_{\max} \propto 1/T$$

where T is the natural period of the system. Codes of practice for seismic loading reflect this trend of reducing lateral load coupled with lengthening natural period.

In our example building this particular earthquake loading governs, but only just. For tall buildings wind loading is frequently the governing lateral load.

*Example 4.4   Surge response of a tension-leg platform when heave
is accounted for*

In Example 3.1 the basic dynamic characteristics of a tension-leg platform were
calculated. The period for heave motion was estimated to be 1.5 s, while that for
surge was 64 s. The amplitude of the steady-state surge response to a train of
operating waves of period 11 s, and crest-to-trough height of 12 m, was calculated
to be

$$\Delta_s = 1.8 \text{ m}$$

The static wave force in surge was given as 6×10^4 kN and the total mass of the
platform for surge motion was 1.035×10^5 kN s^2 m^{-1}.

However, when the 6 m crest of the wave goes by, the tension in the legs increases
by about 80%, while a reduction in tension of 80% below the static preset tension
is recorded when the trough goes by. Thus, the equation of motion to be solved,
if due allowance is to be made for the oscillating leg tension and therefore oscillating
stiffness of the platform in surge, is

$$\ddot{\Delta} + 2\zeta\omega\dot{\Delta} + \omega^2(1 + \bar{a} \sin \Omega t)\Delta = (F/M) \sin \Omega t$$

where $\omega = 2\pi/64$ rad s^{-1}, $\zeta = 0.03$, $\bar{a} = 0.8$, $\Omega = 2\pi/11$ rad s^{-1}, and $F/M = 6 \times 10^4/$
1.035×10^5 m s^{-2}.

We seek a steady-state solution to this problem in order to provide a comparison
with the earlier result. The mathematical theory for such a solution, although
complicated, has long been established. However, the numerical scheme below gives
results quite quickly. The solution scheme is, from rest,

$$\Delta_0 = 0$$

$$\Delta_1 = (\tfrac{1}{3}a_0 + \tfrac{1}{6}a_1)\tau^2$$

$$(1 + \zeta\omega\tau)\Delta_k = (\tfrac{1}{6}a_i + \tfrac{2}{3}a_j + \tfrac{1}{6}a_k)\tau^2 + [2 - (\omega\tau)^2(1 + \bar{a} \sin \Omega t_j)]\Delta_j$$
$$- (1 - \zeta\omega\tau)\Delta_i \qquad k = 2, 3, 4, \dots$$

A suitable increment in time is 1 s, which allows the forcing function (of period
11 s) to be accurately portrayed: it is obviously entirely adequate in terms of the
original structural period of 64 s. An adequate steady state is achieved after some
2000 s in this instance, although this could have been hastened by increasing the
damping, which would have had minimal effect on the magnitude of the answers
because the system is so compliant. The result for one cycle of steady-state response
is shown in Figure 4.9.

Notice that the period of response is 11 s, as it should be. Notice further that,
while the positive peak is now significantly different from the negative peak, the
algebraic sum of 3.5 m is not significantly different from the value of $2 \times 1.8 = 3.6$ m
for the system when the fluctuating lateral stiffness is ignored. Because response is
almost 180° out of phase with forcing, the peak positive response amplitude of
2.35 m coincides with the wave trough which, itself, is associated with the lowest
lateral stiffness of the system.

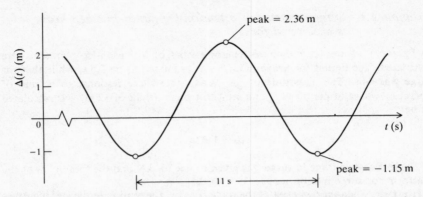

Figure 4.9 Steady-state response of a tension-leg platform in surge when heave response is included.

Conversely, the negative peak in response of 1.15 m is smaller because it corresponds to the crest of the wave and, therefore, to the highest lateral stiffness. Thus, in a local sense, this distinction between positive and negative peaks in response can be explained by the usual arguments relating to force and stiffness once it has been recognised that response is, nonetheless, almost completely out of phase with forcing. This conclusion is not altered by the fact that a net increase in stiffness of the tension-leg platform (say, due to a reduction in water depth) would lead, in the first instance, to a stronger response to the same operating wave. This point was made in Section 3.1.1.

In this example the stiffness is seen to depend fairly directly on the loading level (after all, higher waves give rise to higher wave forces). Because of this, the linearity of the system is no longer maintained and it is no longer true that an increase in forcing leads to a commensurate increase in response. The response is non-linear to some degree.

Example 4.5

This example is a continuation of Problems 3.6 and 3.7 where a tall reinforced concrete tower was analysed for the effects of vortex shedding. It was shown that a resonant response to vortex shedding occurs when the wind velocity is 28 m s^{-1} because then the frequency of vortex shedding is identical to the fundamental natural frequency of the tower of about 2 rad s^{-1}. The tower, of height 200 m and outside diameter 17.5 m, has equivalent mass and equivalent stiffness of $K_e = 13\,400 \text{ kN m}^{-1}$ and $M_e = 3400 \text{ kN s}^2 \text{m}^{-1}$, and the fraction of critical damping is estimated to be $\zeta = 0.03$.

When vortex shedding occurs, which gives rise to cross-wind response, there is also excitation in the along-wind direction. A model for this gusting action, which has been suggested by Davenport (1967), is for a wind velocity of the form

$$V(t) = V + \delta V \sin \Omega t$$

where V is the mean wind velocity, δV is the magnitude of the fluctuating component and Ω is the frequency of vortex shedding in radians per second. Thus the drag force on a cylinder of unit height is

$$C_d \rho \, \frac{V(t)^2}{2} \, D$$

where C_d is the drag coefficient and D is the diameter of the cylinder. Consequently,

$$p(t) = C_d \rho \, \frac{V^2}{2} \, D \left[1 + \frac{2\delta V}{V} \sin \Omega t + \left(\frac{\delta V}{V} \right)^2 \sin^2 \Omega t \right]$$

The first term on the right represents the static wind drag and the fluctuating components occur at a frequency given by the Strouhal number relationship

$$S = \frac{fD}{V} \approx 0.20$$

where V is the mean velocity and $f = \Omega/2\pi$ cycles per second.

In this example we shall assume that the mean velocity is again constant with height at 28 m s^{-1}. This constancy with height is manifestly untrue and one sees that by adopting a more realistic profile the problem of determining dynamic response is further complicated because the local frequency of vortex shedding is dependent on the level of the tower under consideration. Nevertheless, it is often conservative to assume a uniform profile when the mean value is that which gives rise to vortex shedding of a frequency identical to the natural frequency of the tower. As usual, less restrictive assumptions may be made if the answers prove to be unpalatable. For example, this could be forced when the design wind speed is significantly higher than that giving rise to resonance from vortex shedding.

The equation of motion for tower top displacement is

$$\ddot{\Delta} + 2\zeta\omega \, \dot{\Delta} + \omega^2 \, \Delta = \frac{P_e(t)}{M_e}$$

where $\zeta = 0.03$, $\omega = \Omega = 2 \text{ rad s}^{-1}$ and

$$P_e(t) = \int_0^{200} p(t) \left(\frac{z}{200} \right)^2 \, dz$$

Substitution and integration, based on assumed values of $\delta V = 12 \text{ m s}^{-1}$ and $C_d = 0.7$, gives

$$P_e(t) = 385(1 + 0.86 \sin \Omega t + 0.18 \sin^2 \Omega t) \text{ kN}$$

and so the static drift under a velocity of 28 m s^{-1} is

$$\Delta_{st} = \frac{385}{\omega^2 M_e} = \frac{385}{K_e} = \frac{385}{13\,400} = 0.029 \text{ m}$$

137

The peak dynamic displacement at steady state assuming a fluctuating drag force of the simple form

$$385 \times 0.86 \sin \Omega t = 331 \sin \Omega t$$

is

$$\Delta_{max} = \frac{(331/K_e)}{2\zeta} = 0.41 \text{ m}$$

Integration of the equation of motion in the form

$$\ddot{\Delta} + 0.12\dot{\Delta} + 4\Delta = 0.097 \sin 2t + 0.020 \sin^2 2t$$

with a time step of $\tau = \pi/10$ s yields, eventually, a steady-state response with a positive peak of 0.342 m and a negative peak of -0.337 m. If, to this, is added the static response of 0.029 m, the positive and negative amplitudes of the along-wind response are 0.37 m and -0.31 m, respectively.

Therefore, given the along-wind model chosen, and assuming a fluctuating velocity component of 12 m s^{-1}, the tower top will undergo a circular motion in plan of amplitude 0.34 m, approximately (recall that the cross-wind response was shown to be of amplitude 0.34 m, approximately).

The peak displacement, being only about 1/600 of the height, forces one to conclude that the tower is adequately stiff in both the static and dynamic senses.

4.4 Numerical solution with material non-linearity

In this section we discuss the development of a simple numerical routine that permits dynamic analysis when the system is idealised as an elasto-plastic oscillator. That is, there is a clearly defined limit to the elastic capacity of the structure, and once this internal force level is reached it holds steady at that level. After this development the technique is then applied to an example and certain comparisons are made with response based on the premise of infinite elasticity.

Figure 4.10 shows the idealised resistance–deflection relationship assumed. Response is assumed to start from rest, namely position 0 in the figure. Note that the limiting elastic displacement Δ_{el} is equal to R/K where R is the resistance supplied. The resistance level is sometimes assumed to be independent of the stiffness K but this cannot, in reality, be so.

In starting from rest at 0 there will always be an initial elastic portion governed by the approximate equations

$$\Delta_0 = 0$$

$$\Delta_1 = (\tfrac{1}{3}a_0 + \tfrac{1}{6}a_1)\tau^2$$

$$\Delta_k = (\tfrac{1}{6}a_i + \tfrac{2}{3}a_j + \tfrac{1}{6}a_k)\tau^2 + [2 - (\omega\tau)^2]\Delta_j - \Delta_i \qquad k = 2, 3, 4, \ldots \qquad (4.33)$$

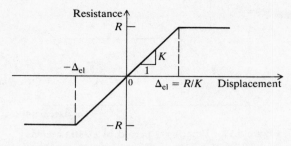

Figure 4.10 Definition diagram for elasto-plastic response.

Damping may be ignored as it is generally insignificant when inelastic behaviour of any consequence is possible.

Response will either remain elastic − that is $\pm \Delta_{el}$ is never exceeded − in which case the above equations are fine, or it will enter the inelastic region. Suppose Δ_k becomes such that Δ_{el} is just exceeded and suppose, for convenience, that we are in the positive quadrant of Figure 4.10. Let us call this displacement Δ_J: it is shown in Figure 4.11.

The displacement Δ_I (which is slightly less than Δ_{el}) is also shown. We wish now to move out along the horizontal tail of the resistance−deflection curve and to do this we need to correct the spring force, which is too high when equilibrium is being written for step J.

As can be seen from Figure 4.11 the force to be subtracted is

$$K(\Delta_J - \Delta_{el})$$

Our difference formulation is thus

$$(\Delta_K - 2\,\Delta_J + \Delta_I)/\tau^2 + \omega^2\,\Delta_J - \omega^2(\Delta_J - \Delta_{el}) = (\tfrac{1}{6}a_I + \tfrac{2}{3}a_J + \tfrac{1}{6}a_K)$$

which becomes

$$\Delta_K = (\tfrac{1}{6}a_I + \tfrac{2}{3}a_J + \tfrac{1}{6}a_K)\tau^2 + 2\Delta_J - \Delta_I - (\omega\tau)^2\Delta_{el}$$

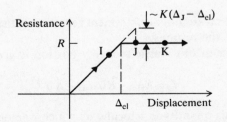

Figure 4.11 Entering the plastic region of response.

139

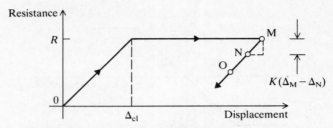

Figure 4.12 Entering a period of elastic recovery.

In this way the spring force is held constant at R ($= K\Delta_{el}$), which is as required. So, the general equation for moving out along the horizontal tail is

$$\Delta_k = (\tfrac{1}{6}a_i + \tfrac{2}{3}a_j + \tfrac{1}{6}a_k)\tau^2 + 2\Delta_j - \Delta_i - (\omega\tau)^2\Delta_{el} \qquad (4.34)$$

The response as calculated above is non-linear.

This equation is used until the peak inelastic displacement Δ_M is recorded. This may be isolated because Δ_N, the next displacement calculated by Equation 4.34, is smaller than Δ_M. At this point the equation changes again and we enter a period of elastic recovery as depicted by Figure 4.12.

A force $K(\Delta_M - \Delta_N)$ must be subtracted from the resistance R ($= K\Delta_{el}$) when writing equilibrium at step N, namely

$$(\Delta_O - 2\,\Delta_N + \Delta_M)/\tau^2 + \omega^2\Delta_{el} - \omega^2(\Delta_M - \Delta_N) = (\tfrac{1}{6}a_M + \tfrac{2}{3}a_N + \tfrac{1}{6}a_O)$$

giving

$$\Delta_O = (\tfrac{1}{6}a_M + \tfrac{2}{3}a_N + \tfrac{1}{6}a_O)\tau^2 + [2 - (\omega\tau)^2]\Delta_N - \Delta_M + (\omega\tau)^2(\Delta_M - \Delta_{el})$$

What is important to realise about this equation is that the quantity

$$(\omega\tau)^2(\Delta_M - \Delta_{el})$$

is a constant for all subsequent calculations in this elastic portion of the time history of response. Therefore in this region our equation is

$$\Delta_k = (\tfrac{1}{6}a_i + \tfrac{2}{3}a_j + \tfrac{1}{6}a_k)\tau^2 + [2 - (\omega\tau)^2]\Delta_j - \Delta_i + (\omega\tau)^2(\Delta_{max} - \Delta_{el}) \qquad (4.35)$$

where Δ_{max} is that maximum displacement recorded at the end of the earlier plastic portion of the response.

The spring force in this elastic recovery portion is given by

$$F_j = K\Delta_{el} - K(\Delta_{max} - \Delta_j) \qquad (4.36)$$

and the size of this quantity is a means by which one decides whether the system is about to enter the positive plastic region (again) or the negative plastic region: either one is possible. In the case of a return to positive yield, that condition arises when

$$F_j = K \Delta_{\text{el}}$$

that is, when

$$\Delta_j = \Delta_{\text{max}} \qquad\qquad (4.37)$$

Because of the discrete nature of the problem it is impossible to achieve this equality exactly and so, inevitably, one goes one step past, calculating Δ_k, and then changing to Equation 4.34 to continue on another plastic excursion.

A numerical routine has to keep open the above eventuality, of course, but it is of more interest in the present scheme of things to consider the situation when negative yield is approached, that is, when

$$F_j = -K \Delta_{\text{el}} = K\Delta_{\text{el}} - K(\Delta_{\text{max}} - \Delta_j)$$

or

$$\Delta_j = \Delta_{\text{max}} - 2\Delta_{\text{el}} \qquad\qquad (4.38)$$

It is again apparent that one cannot exactly meet this requirement and the spring force will overshoot.

In Figure 4.13 the displacement Δ_Q depicts this overshoot and is the last displacement that can be calculated by Equation 4.35. To calculate Δ_R we must add a quantity

$$K[(\Delta_{\text{max}} - 2\Delta_{\text{el}}) - \Delta_Q]$$

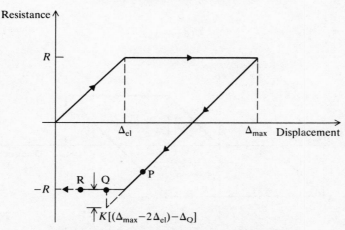

Figure 4.13 Entering the negative yield region.

to the spring force term in the equation of motion that governed the elastic recovery portion. The correction is to be added because we are correcting for negative yield. Thus

$$\Delta_R = (\tfrac{1}{6}a_P + \tfrac{2}{3}a_Q + \tfrac{1}{6}a_R)\tau^2 + [2 - (\omega\tau)^2]\Delta_Q - \Delta_P$$

$$+ (\omega\tau)^2 (\Delta_{max} - \Delta_{el}) - (\omega\tau)^2 [(\Delta_{max} - \Delta_{el}) - \Delta_Q]$$

That is

$$\Delta_R = (\tfrac{1}{6}a_P + \tfrac{2}{3}a_Q + \tfrac{1}{6}a_R)\tau^2 + 2\Delta_Q - \Delta_P + (\omega\tau)^2\Delta_{el}$$

and in proceeding along the horizontal tail of the hysteresis loop towards the origin the formulation takes the form

$$\Delta_k = (\tfrac{1}{6}a_i + \tfrac{2}{3}a_j + \tfrac{1}{6}a_k)\tau^2 + 2\Delta_j - \Delta_i + (\omega\tau)^2\Delta_{el} \qquad (4.39)$$

Evidently, the only difference between this and Equation 4.34 – which applies to the upper horizontal tail – is a change of sign in front of the term $(\omega\tau)^2\Delta_{el}$. The physical reason is simply that the yield force is negative now, instead of positive, as before.

Proceeding, one eventually reaches a point at which the vibrational velocity is again momentarily zero and a point of minimum displacement has been reached. This minimum displacement may be either negative or positive. Figure 4.14 is for the case where it is negative. We then begin another period of elastic recovery.

The displacement Δ_S is the minimum in a local sense for that part of the hysteresis loop. The correction required to calculate Δ_U is to add a force increment

$$K(\Delta_T - \Delta_S)$$

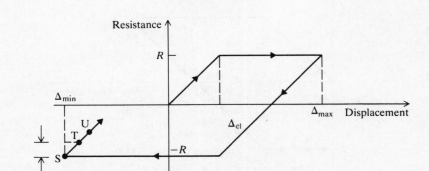

Figure 4.14 Elastic recovery from negative yield.

to the negative yield force $-K\Delta_{el}$. This gives

$$(\Delta_U - 2\Delta_T + \Delta_S)/\tau^2 - \omega^2\Delta_{el} + \omega^2(\Delta_T - \Delta_S) = (\tfrac{1}{6}a_S + \tfrac{2}{3}a_T + \tfrac{1}{6}a_U)$$

That is

$$\Delta_U = (\tfrac{1}{6}a_S + \tfrac{2}{3}a_T + \tfrac{1}{6}a_U)\tau^2 + [2 - (\omega\tau)^2]\Delta_T - \Delta_S + (\omega\tau)^2(\Delta_S + \Delta_{el})$$

However, this last term stays constant and it is better to write it as

$$(\omega\tau)^2(\Delta_{min} + \Delta_{el})$$

in order to establish the equation governing elastic recovery from a post-elastic excursion involving negative yield, which is

$$\Delta_k = (\tfrac{1}{6}a_i + \tfrac{2}{3}a_j + \tfrac{1}{6}a_k)\tau^2 + [2 - (\omega\tau)^2]\Delta_j - \Delta_i + (\omega\tau)^2(\Delta_{min} + \Delta_{el}) \quad (4.40)$$

The only difference between this and its 'mirror image' counterpart (Eqn. 4.35) is the change of sign in front of Δ_{el}.

Again, in this region of elastic recovery the spring force is

$$F_j = -K\Delta_{el} + K(\Delta_j - \Delta_{min}) \quad (4.41)$$

The response either continues elastically from here, or it goes plastic again after the spring force has reached negative yield, i.e. when

$$\Delta_j = \Delta_{min} \quad (4.42)$$

or it goes plastic after the spring force has recovered enough to reach positive yield, i.e. when

$$\Delta_j = \Delta_{min} + 2\Delta_{el} \quad (4.43)$$

At this juncture, we have covered all the situations that can arise. In the vast majority of cases involving inelastic calculations, those calculations are best performed by writing a short computer program. The above equations are sufficient for that purpose.

There are a few further points to be made: first, rather small time steps are desirable – much smaller than would be necessary with an elastic analysis; secondly, it is necessary to ensure that whenever a hysteresis loop is being traversed it is traversed in clockwise manner – anticlockwise is physically meaningless; thirdly, the above descriptions cover only the very simplest situation of material non-linearity – routines that allow for stiffness and strength degradation, for example, can and have been written: this permits the results of experimental studies of post-elastic performance of subassemblages, where strength and stiffness both tend to degrade with repeated cycles of post-elastic response, to be incorporated more directly. A great deal of work has been done in this field over the past 30 years.

The above is at best a brief introduction to the field of inelastic dynamic analysis. The following example illustrates some of the practical details in obtaining solutions when the linearity of the response is limited.

Example 4.6 Comparison of elastic and inelastic response for a given oscillator responding to an idealised earthquake

The structure to which the idealised earthquake is applied is a single-storey structure shown in schematic form in Figure 4.15.

The mass $M = 8 \text{ kN s}^2 \text{ m}^{-1}$, the total stiffness is $K = 900 \text{ kN m}^{-1}$, giving the natural period as

$$T = 2\pi \sqrt{(M/K)} \approx 0.6 \text{ s}$$

The input record is given at intervals of 0.01 s, which is far shorter than it need be insofar as the period of vibration is concerned. Nevertheless, we take a value of $\tau = 0.01$ s as the integration time step, leading to a value of $\omega\tau = 0.106$ and $(\omega\tau)^2 = 0.0113$.

The input record consists of a total of 3 s of strong motion with a peak ground acceleration of 30% g. The input acceleration is listed in Table 4.2. The purpose of the example is twofold: first, an elastic analysis is run; secondly, a limit is placed on the lateral strength of the structure and inelastic behaviour is permitted – this allows comparisons to be made and, indeed, tentative conclusions to be drawn.

ELASTIC ANALYSIS

The equation of motion is

$$\ddot{\Delta} + \omega^2\Delta = -a(t)$$

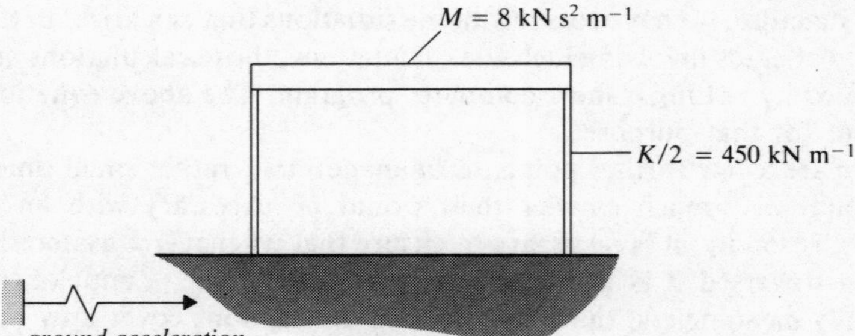

$M = 8 \text{ kN s}^2 \text{ m}^{-1}$

$K/2 = 450 \text{ kN m}^{-1}$

ground acceleration

Figure 4.15 Schematic for a single-storey structure responding to earthquake ground motion.

where $-a(t)$ is a piecewise linear record of an idealised earthquake. Damping is ignored. Our solution routine is

$$\Delta_0 = 0$$

$$\Delta_1 = (\tfrac{1}{3}a_0 + \tfrac{1}{6}a_1)0.0001$$

$$\Delta_k = (\tfrac{1}{6}a_i + \tfrac{2}{3}a_j + \tfrac{1}{6}a_k)0.0001 + 1.9887\Delta_j - \Delta_i \qquad k = 2, 3, 4, \ldots$$

for a time step of 0.01 s.

Partial results are shown in Table 4.2a; that is, results are given at every 0.05 s throughout the motion. The peak displacement is 0.0597 m, occurring 1 s after the start. Multiplying this by the spring constant gives a peak base shear of

$$S_{max} = 900 \times 0.0597 = 53.7 \text{ kN}$$

Comparing this with the weight of the structure, which is 78.5 kN, it is seen that the peak base shear, as a fraction of the weight of the structure, is

$$\frac{S_{max}}{Mg} = \frac{53.7}{78.5} = 0.684$$

The peak ground acceleration was $0.3g$ so that the peak response is more than twice as great. This structure, therefore, responds strongly to this 'earthquake'.

INELASTIC RESPONSE

Suppose that the lateral strength supplied to the structure is limited, rather than unlimited as implied above. Suppose, therefore, that the level of lateral resistance built in is $20\% g$. That is

$$R = 0.2Mg = 16 \text{ kN}$$

The lateral displacement at the elastic limit is

$$\Delta_{el} = \frac{R}{K} = \frac{16}{900} = 0.0178 \text{ m}$$

Response is shown in Table 4.2b.

The elastic and inelastic regions are delineated and sufficient information is presented to allow displacement values to be checked using the earlier formulae (Eqns 4.34–40). Printout of the spring force provides a useful way of keeping a check on which portion of the hysteresis loop one is on and this feature is commended to the reader. The dots indicate regions where the information has been deleted simply to save space. Figure 4.16 shows the same information plotted as a series of parallelograms in the form of hysteresis loops. The net area under such load–deflection curves gives a measure of the energy dissipated by plastic action.

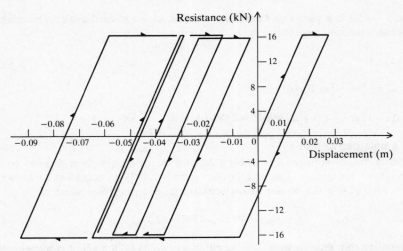

Figure 4.16 Hysteresis loop for single-storey structure responding in elasto-plastic fashion to an earthquake loading.

The permanent set of the structure after the earthquake has stopped is approximately -0.0466 m (the value at the end of the run), to which is added 0.0178 m (representing elastic recovery), namely -0.029 m. Notice that the peak displacement is $\Delta'_{max} = -0.0926$ m. In terms of the permissible elastic displacement, this represents a ductility factor of

$$\mu = \frac{\Delta'_{max}}{\Delta_{el}} = \frac{0.0926}{0.0178} = 5.20$$

This result, which is broadly typical of what one may expect from structures with these characteristics responding to similar levels of strong motion input, gives rise to the following comments.

For the structure considered as a wholly elastic body the peak strain energy stored is

$$\tfrac{1}{2}K\Delta^2_{max} = \tfrac{1}{2}F^2_{max}/K$$

where F_{max} is the peak elastic spring force. The amount of energy taken to traverse, monotonically, the elasto-plastic response curve shown in Figure 2.20 is

$$R(\Delta'_{max} - \Delta_{el}/2) = (R^2/2K)(2\mu - 1)$$

where $\mu = \Delta'_{max}/\Delta_{el}$ is the ductility factor and R is the resistance level.

As a result of numerous studies conducted on the inelastic response of structures to earthquakes, it has been found that, for structures whose periods lie in the range of, say, 0.5 s to 2.5 s, a good estimate of the peak ductility demand of a structure can be found by making the assumption that the two energy quantities, above, are equal. Thus (look at Eqn. 2.26)

$$R/F_{max} = 1/\sqrt{(2\mu - 1)}$$

Consider our evidence: From our earlier elastic analysis

$$F_{max} = 900 \times 0.0597 = 53.7 \text{ kN}$$

From our inelastic analysis, where we chose $R = 16$ kN, we found $\mu = 5.2$. Hence

$$R/F_{max} = 16/53.7 = 0.298$$

whereas

$$1/\sqrt{(2\mu - 1)} = 0.326$$

Clearly, agreement is reasonable in this case, which lends support to the general contention.

The implication of this is that inelastic analyses need not necessarily be carried out in order to estimate response in the inelastic region. Rather, knowing the likely elastic response, and having decided, *a priori*, on the level of ductility to be 'detailed' into the structure, one may determine the lateral resistance to be supplied. This, in fact, is the basis behind all modern codes of practice for earthquake loading where, for ductile structures, the suggested resistance levels are often far below the levels that would be required if the structure remained elastic.

For further information in the general area of earthquake engineering, the book by Dowrick (1977) and the continuing series of monographs written by distinguished earthquake engineers of the Earthquake Engineering Research Institute (Chopra 1981, Housner & Jennings 1982, Newmark & Hall 1982) are useful.

Table 4.2 Response to an idealised earthquake.

(a) Elastic case

Time (s)	Displacement (m)	Input accn (m s^{-2})
0.00	0.00000	0.00
0.05	0.00016	0.40
0.10	0.00126	0.80
0.15	0.00396	1.20
0.20	0.00851	1.60
0.25	0.01462	2.00
0.30	0.02119	1.40
0.35	0.02536	0.80
0.40	0.02451	0.20
0.45	0.01742	−0.40
0.50	0.00456	−1.00
0.55	−0.01195	−1.50
0.60	−0.02884	−2.00
0.65	−0.04257	−2.20
0.70	−0.04948	−1.20

Table 4.2 continued

Time (s)	Displacement (m)	Input accn (m s^{-2})
0.75	− 0.04597	− 0.80
0.80	− 0.03138	0.60
0.85	− 0.00711	1.00
0.90	0.02155	1.40
0.95	0.04697	0.00
1.00	0.05966	− 1.00
1.05	0.05425	− 0.20
1.10	0.03336	0.40
1.15	0.00388	0.00
1.20	− 0.02602	1.20
1.25	− 0.04616	1.60
1.30	− 0.05095	− 1.00
1.35	− 0.04352	− 2.00
1.40	− 0.02902	− 3.00
1.45	− 0.01285	− 1.50
1.50	0.00323	0.10
1.55	0.01822	0.60
1.60	0.02896	− 0.60
1.65	0.03066	− 0.90
1.70	0.02268	1.10
1.75	0.01074	2.10
1.80	0.00078	2.60
1.85	− 0.00396	0.90
1.90	− 0.00512	− 0.10
1.95	− 0.00458	0.20
2.00	− 0.00238	0.30
2.05	0.00174	1.70
2.10	0.00908	2.00
2.15	0.01828	1.00
2.20	0.02449	− 1.00
2.25	0.02214	− 1.50
2.30	0.01066	− 0.50
2.35	− 0.00488	0.70
2.40	− 0.01766	1.20
2.45	− 0.02262	1.80
2.50	− 0.01697	2.40
2.55	− 0.00194	0.20
2.60	0.01452	− 1.00
2.65	0.02524	− 0.50
2.70	0.02794	0.30
2.75	0.02340	0.40
2.80	0.01325	0.10
2.85	− 0.00063	− 1.00
2.90	− 0.01653	− 1.50
2.95	− 0.03081	− 0.20
3.00	− 0.03757	0.00

(b) Inelastic case

Time (s)	Displacement (m)	Spring force (kN)	Input accn (m s^{-2})
elastic region			
0.00	0.000000	0.0	0.00
0.01	0.000001	0.0	0.08
0.02	0.000011	0.0	0.16
0.03	0.000036	0.0	0.24
0.04	0.000085	0.1	0.32
0.05	0.000164	0.1	0.40
⋮	⋮		⋮
0.26	0.015984	14.4	1.88
0.27	0.017344	15.6	1.76
0.28	0.018685	16.8	1.64
plastic region			
0.29	0.019991	16.0	1.52
0.30	0.021248	16.0	1.40
⋮	⋮	⋮	⋮
0.39	0.028423	16.0	0.32
0.40	0.028540	16.0	0.20
0.41	0.028477	16.0	0.08
elastic region			
0.42	0.028223	15.7	− 0.04
0.43	0.027769	15.3	− 0.16
⋮	⋮	⋮	⋮
0.58	− 0.002081	− 11.6	− 1.80
0.59	− 0.005067	− 14.2	− 1.90
0.60	− 0.008065	− 16.9	− 2.00
plastic region			
0.61	− 0.011062	− 16.0	− 2.04
0.62	− 0.014063	− 16.0	− 2.08
⋮	⋮		⋮
0.84	− 0.065027	− 16.0	0.92
0.85	− 0.065222	− 16.0	1.00
0.86	− 0.065117	− 16.0	1.08
elastic region			
0.87	− 0.064705	− 15.5	1.16
0.88	− 0.063983	− 14.9	1.24
⋮	⋮	⋮	⋮
1.07	− 0.030558	15.2	0.04
1.08	− 0.029973	15.7	0.16
1.09	− 0.029569	16.1	0.28
plastic region			
1.10	− 0.029337	16.0	0.40
1.11	− 0.029268	16.0	0.32
1.12	− 0.029367	16.0	0.24

Table 4.2 continued

Time (s)	Displacement (m)	Spring force (kN)	Input accn (m s^{-2})
elastic region			
1.13	−0.029641	15.7	0.16
1.14	−0.030095	15.3	0.08
⋮	⋮	⋮	⋮
1.39	−0.060268	−11.9	−2.80
1.40	−0.062512	−13.9	−3.00
1.41	−0.064874	−16.0	−2.70
plastic region			
1.42	−0.067307	−16.0	−2.40
1.43	−0.069779	−16.0	−2.10
⋮	⋮	⋮	⋮
1.57	−0.092517	−16.0	0.12
1.58	−0.092560	−16.0	−0.12
1.59	−0.092415	−16.0	−0.36
elastic region			
1.60	−0.092108	−15.6	−0.60
1.61	−0.091663	−15.2	−0.66
⋮	⋮	⋮	⋮
1.79	−0.060899	12.5	2.50
1.80	−0.057723	15.4	2.60
1.81	−0.054486	18.3	2.26
plastic region			
1.82	−0.051224	16.0	1.92
1.83	−0.047969	16.0	1.58
⋮	⋮	⋮	⋮
2.03	−0.014014	16.0	1.14
2.04	−0.014000	16.0	1.42
2.05	−0.014043	16.0	1.70
elastic region			
2.06	−0.014119	15.9	1.76
2.07	−0.014218	15.8	1.82
⋮	⋮	⋮	⋮
2.32	−0.045712	−12.5	−0.02
2.33	−0.048097	−14.7	0.22
2.34	−0.050277	−16.7	0.46
plastic region			
2.35	−0.052211	−16.0	0.70
2.36	−0.053878	−16.0	0.80
⋮	⋮	⋮	⋮
2.40	−0.057644	−16.0	1.20
2.41	−0.057810	−16.0	1.32
2.42	−0.057644	−16.0	1.44
elastic region			
2.43	−0.057136	−15.4	1.56
2.44	−0.056279	−14.6	1.68
⋮	⋮	⋮	⋮

Time (s)	Displacement (m)	Spring force (kN)	Input accn $(m\,s^{-2})$
2.55	− 0.027063	11.7	0.20
2.56	− 0.023724	14.7	− 0.04
2.57	− 0.020574	17.5	− 0.28
plastic region			
2.58	− 0.017651	16.0	− 0.52
2.59	− 0.014980	16.0	− 0.76
⋮	⋮	⋮	⋮
2.68	− 0.003314	16.0	− 0.02
2.69	− 0.003217	16.0	0.14
2.70	− 0.003306	16.0	0.30
elastic region			
2.71	− 0.003565	15.7	0.32
2.72	− 0.003989	15.3	0.34
⋮	⋮	⋮	⋮
2.92	− 0.036181	− 13.7	0.00
2.93	− 0.038176	− 15.5	0.00
2.94	− 0.039978	− 17.1	0.00
plastic region			
2.95	− 0.041580	− 16.0	0.00
2.96	− 0.042982	− 16.0	0.00
2.97	− 0.044184	− 16.0	0.00
2.98	− 0.045186	− 16.0	0.00
2.99	− 0.045988	− 16.0	0.00
3.00	− 0.046590	− 16.0	0.00

4.5 Computer program for linear dynamic analysis

A FORTRAN 77 listing of a program to calculate response displacements, velocities and accelerations is given. The input consists of the time step increment, the natural circular frequency of the system, the fraction of critical damping, the input acceleration at each time step and the total number of time steps. The program is based on the recursive relationship, Equation 4.29.

Sample output, based on Example 4.1, is also included.

151

DUHAMEL'S INTEGRAL

```
      program dyana(input=80,output=140,tape5=input,tape6=output)
c
c
c  TIME HISTORY ANALYSIS FOR A SINGLE DEGREE OF FREEDOM OSCILLATOR
c
c
c  ********************************************************************
c
c  housekeeping of dyana
c
c
c  a     = vector of input accelerations at each time step (m/sec/sec)
c  d     = vector of displacements (m)
c  vel   = vector of velocities (m/sec)
c  acc   = vector of accelerations (m/sec/sec)
c  tau   = time step increment (sec)
c  zeta  = fraction of critical damping
c  omega = natural frequency (rad/sec)
c  nstep = total number of time steps
c
c  ********************************************************************
c
c
      dimension d(3000),a(3000),vel(3000),acc(3000),title(8)
      data nr,nw/5,6/
      read(nr,100) title
  100 format(8a10)
      write(nw,200) title
  200 format(1h1,8a10)
c
c  input is in free format
c
c  input data
c  max. no. of time steps : 3000
c
      read(nr,*) nstep,tau,zeta,omega
      write(nw,300) nstep,tau,zeta,omega
  300 format(//1x,' no. of time steps =',i5,/1x,' time step increment =
     *',1pe13.4,'  (sec)'/1x,' fraction of critical damping =',1pe13.4,/
     *1x,' natural frequency =',1pe13.4,'  (rad/sec)')
c
      t=0.0
      do 1000 i=1,3000
      a(i)=0.579*sin(0.571*t)
      t=t+1
 1000 continue
c
      write(nw,400)
  400 format(///1x,' input acceleration :')
      write(nw,500) (a(i),i=1,nstep)
  500 format(//1x,1p8e13.5/(1x,1p8e13.5))
c
c  initialise variables
c
      d(1)=0.0
      tsum=0.0
c
      c1=(1.0/6.0)*tau*tau
      c2=omega*omega
      c3=omega*zeta
      c4=1.0+tau*c3
      c5=(2.0-c2*tau*tau)/c4
      c6=(1.0-tau*c3)/c4
      c7=2.0*tau
      c3=2.0*c3
c
      d(2)=(2.0*a(1)+a(2))*c1
      c1=c1/c4
```

152

LINEAR DYNAMIC ANALYSIS

```
c
c  set up heading for output table
c
      write(nw,600)
  600 format(////1x,'  time',5x,'displacement',5x,' velocity',5x,'acceler
     *ation'//)
      write(nw,620)
  620 format(1x,'    sec',5x,'    m     ',5x,'  m/sec ',5x,' m/sec/s
     *ec')
      write(nw,630)
  630 format(1x,'-------------------------------------------------------
     *'/)
c
c  main loop
c
      do 700 k=3,nstep
c
c  calculate displacement
c
      c5=c5*(1.0+0.8*a(k-1)/0.579)
      d(k)=(a(k-2)+4.0*a(k-1)+a(k))*c1+c5*d(k-1)-c6*d(k-2)
c
c  calculate velocity
c
      vel(k-1)=(d(k)-d(k-2))/c7
c
c  calculate acceleration
c
      acc(k-1)=a(k-1)-c2*d(k-1)-c3*vel(k-1)
c
      tsum=tau+tsum
c
      write(nw,650) tsum,d(k-1),vel(k-1),acc(k-1)
  650 format(1x,f6.1,1pe17.2,1pe14.2,1pe17.2)
  700 continue
c
      tsum=tsum+tau
c
      write(nw,800) tsum,d(nstep)
  800 format(1x,f6.1,1pe17.2)
c
c  end of dyana
c
      stop
      end
```

EXAMPLE 4.1

```
NO. OF TIME STEPS =    7
TIME STEP INCREMENT =   2.0000E-01  (SEC)
FRACTION OF CRITICAL DAMPING =  0.
MATERIAL FREQUENCY =  2.6180E+00  (RAD/SEC)


INPUT ACCELERATION :

0.        5.00000E-01   8.66000E-01   1.00000E+00  8.66000E-01  5.00000E-01   0
```

TIME	DISPLACEMENT	VELOCITY	ACCELERATION
SEC	M	M/SEC	M/SEC/SEC
.2	3.33E-03	6.21E-02	4.77E-01
.4	2.48E-02	1.73E-01	6.96E-01
.6	7.27E-02	2.85E-01	5.02E-01
.8	1.39E-01	3.18E-01	-8.50E-02
1.0	2.00E-01	2.16E-01	-8.70E-01
1.2	2.25E-01		

Problems

4.1 Solve the undamped oscillator equation for a system of period 1 s when the input acceleration is given by a rectangular block with dimensions of $a = 1\ \mathrm{m\,s^{-2}}$ and $t_* = 0.5$ s, in order to find the peak displacement. Refer to Figure 2.16.

4.2 For an input that is a series of steps, show that the recursion formula for the solution to

$$\ddot{\Delta}(t) = a(t)$$

is given by Equation 4.31.

4.3 The exact equation of motion for free vibrations of a simple pendulum is

$$\ddot{\theta} + (2\pi)^2 \sin\theta = 0$$

where θ is the angle of inclination of the pendulum bob to the vertical and the natural circular frequency for small vibration (i.e. $\theta \ll 1$, so that $\sin\theta \approx \theta$) is $\omega = 2\pi\ \mathrm{rad\,s^{-1}}$.

(a) Set up a numerical solution routine to solve the free vibration problem of $\theta(0) = \theta_0$, $\dot{\theta}(0) = 0$.

(b) Hence find the period of vibration for free vibrations in which θ is not vanishingly small. For example, find the period when $\theta_0 = \pi/4$, $\pi/2$ and compare the results.

4.4 This problem represents a continuation of Example 1.3 wherein a 25-storey building was analysed for wind effects. Consider now an idealised gust loading in the form of a suddenly applied triangular load block (see Fig. 2.18). The time of loading, t_*, is 2 s and the peak intensity, per unit height of the frame, is constant with height at $4\ \mathrm{kN\,m^{-1}}$. Use a numerical scheme to find the peak displacement response of the top of the frame. Hence find the dynamic amplification factor for this loading situation and compare the result with that from Figure 2.18.

4.5 Write a computer program to analyse the response of an elasto-plastic oscillator responding to input acceleration. Use the equations of Section 4.4 as a basis for the program and use Example 4.6 for testing purposes.

4.6 A simple oscillator has a mass of $10\ \mathrm{kN\,s^2\,m^{-1}}$, an elastic stiffness of $1000\ \mathrm{kN\,m^{-1}}$ and an elasto-plastic resistance of 12 kN. It is subjected to a suddenly applied load of value 10 kN. Find the peak displacement response and the time at which this peak displacement occurs.

154

5

Multi-degree-of-freedom systems

An introduction

The purpose of this final chapter is to introduce the reader to multi-degree-of-freedom systems, i.e. those structural models in which it is deemed prudent to prescribe two or more independent displacements in order better to describe dynamic response. While it is often true that a carefully constructed single-degree model will capture most features of engineering significance – and this is the emphasis that has been deliberately chosen for this book for practitioners – there are occasions when the discontinuity in mass and/or stiffness is such that a more detailed examination is advisable from the outset. This point, which was made in Chapter 1, could be illustrated as follows: a nuclear reactor containment structure that is a shell of revolution is quite stiff. If it is socketed directly in bedrock of good quality then a single-degree-of-freedom model might well be an appropriate starting point if, say, response to strong ground motion needs to be considered. However, if the structure is founded on soil—bedrock being out of reach, say – the fundamental period of the structure will be much longer because the supporting material is compliant. So, in this instance, in order to account better for the probably significant impact of the interaction of soil and structure, the scope of the model would be widened to include, for example, the rotation of the base, in addition to the lateral movement of the containment.

We shall largely confine our attention to two-degree-of-freedom (2 DOF) systems and will work through the theory illustrating salient features by recourse to a simple model. This seems the best way to proceed since all necessary aspects are covered, and the mathematical manipulations can be done by hand. However, during this treatment an outline will be given of matrix analysis methods for multi-degree systems.

As before, a lengthy example will be presented. In this case it concerns an offshore crane in hoisting mode and it will be modelled as a 2 DOF system – the second degree of freedom being the rigid-body rotation of the boom set in train as a result of the compliance of the luff, or backstay.

155

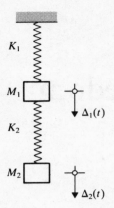

Figure 5.1 Definition diagram for a 2 DOF system.

5.1 Equations of motion for free vibration for a two-degree-of-freedom system

In Figure 5.1 each mass is assigned an absolute displacement (to be measured from the rest position) and is connected by springs, as shown. When the system is disturbed from rest, vibration ensues and our first task is to obtain the equations that govern the dynamic equilibrium of the system. Figure 5.2 shows free-body diagrams for each mass, the dotted arrows being the inertia forces generated by the motion.

For mass 1

$$M_1\ddot{\Delta}_1 + K_1\Delta_1 - K_2(\Delta_2 - \Delta_1) = 0$$

or   (5.1)

$$M_1\ddot{\Delta}_1 + (K_1 + K_2)\Delta_1 - K_2\Delta_2 = 0$$

and for mass 2

$$M_2\ddot{\Delta}_2 + K_2(\Delta_2 - \Delta_1) = 0$$

or   (5.2)

$$M_2\ddot{\Delta}_2 - K_2\Delta_1 + K_2\Delta_2 = 0$$

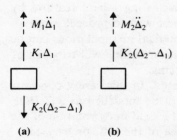

(a)   (b)   Figure 5.2 Freebody diagram of the equilibrium.

These equations of motion may be put in matrix form as

$$\begin{bmatrix} M_1 & 0 \\ 0 & M_2 \end{bmatrix} \begin{Bmatrix} \ddot{\Delta}_1 \\ \ddot{\Delta}_2 \end{Bmatrix} + \begin{bmatrix} (K_1 + K_2) & -K_2 \\ -K_2 & K_2 \end{bmatrix} \begin{Bmatrix} \Delta_1 \\ \Delta_2 \end{Bmatrix} = \begin{Bmatrix} 0 \\ 0 \end{Bmatrix} \tag{5.3}$$

Both matrices are symmetric, as is always the case in such problems. Notice, in addition, that the mass matrix is diagonal: this is a direct consequence of choosing Δ_1 and Δ_2 as absolute displacement quantities. However, the stiffness matrix is coupled. The phrase 'statically coupled' is used to describe a set of equations such as Equations 5.3. If, however, we had chosen Δ_2 as a displacement relative to Δ_1 (Δ_1 being unchanged since it is relative to the fixed support anyway), the equations of motion would have read

$$\begin{bmatrix} (M_1 + M_2) & M_2 \\ M_2 & M_2 \end{bmatrix} \begin{Bmatrix} \ddot{\Delta}_1 \\ \ddot{\Delta}_2 \end{Bmatrix} + \begin{bmatrix} K_1 & 0 \\ 0 & K_2 \end{bmatrix} \begin{Bmatrix} \Delta_1 \\ \Delta_2 \end{Bmatrix} = \begin{Bmatrix} 0 \\ 0 \end{Bmatrix} \tag{5.4}$$

This system of equations is dynamically coupled − the stiffness matrix is now diagonal − but there is no difference of any significance, in that, whatever displacement convention is chosen, the fundamental dynamic characteristics are unaltered. For example, the natural frequencies are the same in either convention since they depend solely on the values of M_1, M_2, K_1 and K_2 pertinent to the particular problem.

The most general case of a 2 DOF system occurs when both matrices are full, as in

$$\begin{bmatrix} M_{11} & M_{12} \\ M_{21} & M_{22} \end{bmatrix} \begin{Bmatrix} \ddot{\Delta}_1 \\ \ddot{\Delta}_2 \end{Bmatrix} + \begin{bmatrix} K_{11} & K_{12} \\ K_{21} & K_{22} \end{bmatrix} \begin{Bmatrix} \Delta_1 \\ \Delta_2 \end{Bmatrix} = \begin{Bmatrix} 0 \\ 0 \end{Bmatrix} \tag{5.5}$$

where, on account of symmetry, $M_{12} = M_{21}$ and $K_{12} = K_{21}$. Such situations are rare − especially for large systems where mass and stiffness matrices are usually sparse − and, in any event, our system of equations (Eqn 5.3), relating as they do to a particular arrangement of masses and springs, is sufficient to permit portrayal of the important features of the response of multi-degree-of-freedom systems.

5.2 Use of Rayleigh's method to determine the approximate value of the fundamental natural frequency

With respect to Figure 5.1, suppose that gravity is acting and that the static response to forces M_1g and M_2g is required. We have, from Equations 5.3

$$\begin{bmatrix} (K_1 + K_2) & -K_2 \\ -K_2 & K_2 \end{bmatrix} \begin{Bmatrix} \Delta_1 \\ \Delta_2 \end{Bmatrix} = \begin{Bmatrix} M_1g \\ M_2g \end{Bmatrix} \tag{5.6}$$

the solution of which is

$$\Delta_1 = \frac{M_1 g + M_2 g}{K_1}$$

(5.7)

$$\Delta_2 = \frac{M_1 g + M_2 g}{K_1} + \frac{M_2 g}{K_2}$$

The ratio of these two displacements is useful in that it allows an estimate to be made of the fundamental natural frequency based on a mode shape which preserves this ratio. Accordingly, consider a fundamental mode given by

$$a_{21} = 1$$

(5.8)

$$a_{11} = \alpha$$

where

$$\alpha = \left(\frac{M_1 g + M_2 g}{K_1}\right) \Big/ \left(\frac{M_1 g + M_2 g}{K_1} + \frac{M_2 g}{K_2}\right)$$

The second subscript refers to the mode, the first refers to the component of the mode associated with the mass.

The peak strain energy stored in the pair of springs when vibration occurs in this assumed fundamental mode is

$$\text{PE} = \tfrac{1}{2} K_1 a_{11}^2 + \tfrac{1}{2} K_2 (a_{21} - a_{11})^2$$

or

(5.9)

$$\text{PE} = \tfrac{1}{2} K_1 \alpha^2 + \tfrac{1}{2} K_2 (1 - \alpha)^2$$

At another phase in the vibration the peak kinetic energy acquired by the two masses is

$$\text{KE} = \tfrac{1}{2} M_1 (\omega_1 a_{11})^2 + \tfrac{1}{2} M_2 (\omega_1 a_{21})^2$$

or

(5.10)

$$\text{KE} = \omega_1^2 (\tfrac{1}{2} M_1 \alpha^2 + \tfrac{1}{2} M_2)$$

Recall that the peak velocity of mass M_1, for example, is $\omega_1 a_{11}$ given that its peak displacement in that mode is a_{11}.

158

Rayleigh's theorem states that the true natural frequency of the first mode, ω_1, is given by

$$\omega_1^2 \leqslant \frac{\frac{1}{2}K_1\alpha^2 + \frac{1}{2}K_2(1-\alpha)^2}{\frac{1}{2}M_1\alpha^2 + \frac{1}{2}M_2}$$

or   (5.11)

$$\omega_1^2 \leqslant \frac{K_1\alpha^2 + K_2(1-\alpha)^2}{M_1\alpha^2 + M_2}$$

and the equality holds only if mode shape chosen is the correct one. Therefore, for an assumed first mode shape, as above, an estimate of the fundamental natural frequency is

$$\omega_1^2 = \frac{K_1\alpha^2 + K_2(1-\alpha)^2}{M_1\alpha^2 + M_2}$$ (5.12)

yielding an answer which will, in general, be too high.

Expressions such as Equation 5.12 can be generalised (see Sec. 5.5) for determination of the fundamental frequency of discrete systems with numerous masses and interconnecting springs. In the limit, when many masses and springs are deployed, one arrives at integral expressions for first-mode quantities like those devised for the examples worked, and problems set, in earlier chapters.

Example 5.1

With reference to Figure 5.1, suppose that $K_1 = 3K$, $K_2 = 2K$ and $M_1 = M_2 = M$ (see Fig. 5.3). Assuming, as was suggested, a mode shape based on spring displacements under gravity, yields, from Equation 5.8

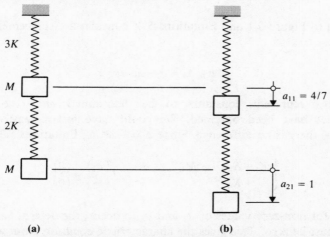

(a)   (b)

Figure 5.3 (a) Model and (b) assumed fundamental mode of Example 5.1.

$$a_{21} = 1$$

$$a_{11} = \alpha = \tfrac{2}{3}/(\tfrac{2}{3} + \tfrac{1}{2}) = \tfrac{4}{7}$$

Hence from Equation 5.12

$$\omega_1^2 = \frac{3 \times (\tfrac{4}{7})^2 + 2 \times (\tfrac{3}{7})^2}{(\tfrac{4}{7})^2 + 1} \frac{K}{M}$$

that is

$$\omega_1^2 = \frac{66}{65} \frac{K}{M}$$

The correct answers are $\omega_1^2 = K/M$ and $a_{21} = 1$, $a_{11} = \tfrac{1}{2}$ (see Example 5.2).

Clearly, agreement in respect of the natural frequency is excellent (the error is less than 1%), but an error of 12.5% exists in the associated mode shape. This error can be reduced significantly by an iteration as follows. The inertia force is $-M\ddot{\Delta}$ or $M\omega^2 \Delta$. Given a first estimate of the mode as $a_{21} = 1$, $a_{11} = \tfrac{4}{7}$ we can therefore apply 'static' forces to each mass in the proportion $\tfrac{4}{7} : 1$. When this is done the resulting deflections may be shown to be in the proportion $\tfrac{22}{43} : 1$. Thus, a much better estimate of the first mode is $a_{21} = 1$, $a_{11} = \tfrac{22}{43}$. This is, in essence, Stodola's variation on a theme due originally to Lord Rayleigh.

Note here that we are talking only of approximations to the first mode. Similar approximate methods exist for the calculation of higher modes. See, for example, Biggs (1964), Clough and Penzien (1975), Paz (1980) and Craig (1981) for more detailed discussion. See, also, Section 5.8 herein.

5.3 Natural frequencies and mode shapes for a two-degree-of-freedom system

Referring to Figure 5.1 and Equations 5.3, consider a displacement of the form

$$\Delta_1, \Delta_2 = a_{1,2} \sin \omega t$$

where $a_{1,2}$ represent constants to be determined once the natural frequencies have been obtained. We could have just as easily written $a \cos \omega t$ – there is no difference. Since $\ddot{\Delta} = -\omega^2 \Delta$, Equations 5.3 become

$$\begin{bmatrix} (K_1 + K_2) - \omega^2 M_1 & -K_2 \\ -K_2 & K - \omega^2 M_2 \end{bmatrix} \begin{Bmatrix} a_1 \\ a_2 \end{Bmatrix} = \begin{Bmatrix} 0 \\ 0 \end{Bmatrix} \qquad (5.13)$$

In order for non-zero values of a_1 and a_2 to occur, the determinant of the matrix must be zero. This gives the characteristic equation from which the natural frequencies are found, namely,

$$[(K_1 + K_2) - \omega^2 M_1](K_2 - \omega^2 M_2) - (-K_2)(-K_2) = 0$$

or (5.14)

$$M_1 M_2 \omega^4 - [M_1 K_2 + M_2(K_1 + K_2)]\omega^2 + K_1 K_2 = 0$$

This is a quadratic in ω^2, the roots of which are

$$\omega_{1,2}^2 = \frac{M_1 K_2 + M_2(K_1 + K_2) \mp \sqrt{\{[M_1 K_2 + M_2(K_1 + K_2)]^2 - 4K_1 K_2 M_1 M_2\}}}{2M_1 M_2} \qquad (5.15)$$

The term inside the square root is smaller than that outside so ω_1^2 cannot be negative and, thus, ω_1 cannot be imaginary. Similarly, the term inside the square root cannot be negative since it may be rearranged to

$$\{[M_1 K_2 - M_2(K_1 + K_2)]^2 + 4M_1 M_2 K_2^2\}$$

which is always positive: therefore, neither ω_1 nor ω_2 can be complex. So, two positive real roots exist for the quadratic in ω^2 and it follows that the two natural frequencies may be taken as positive and real: we were justified in assuming a periodic solution given by the circular trigonometric functions.

The fundamental natural frequency is

$$\omega_1 = \left(\frac{M_1 K_2 + M_2(K_1 + K_2) - \sqrt{\{[M_1 K_2 + M_2(K_1 + K_2)]^2 - 4K_1 K_2 M_1 M_2\}}}{2M_1 M_2}\right)^{\frac{1}{2}}$$

and the second natural frequency is (5.16)

$$\omega_2 = \left(\frac{M_1 K_2 + M_2(K_1 + K_2) + \sqrt{\{[M_1 K_2 + M_2(K_1 + K_2)]^2 - 4K_1 K_2 M_1 M_2\}}}{2M_1 M_2}\right)^{\frac{1}{2}}$$

(5.17)

Associated with each natural frequency is a mode shape which is a vector of displacements, one per degree of freedom, and these displacements are constant when compared to one particular value. This relative constancy is the other basic feature − natural frequency and associated mode shape go hand-in-hand.

The first mode is

$$\{a_1\} = \begin{Bmatrix} a_{11} \\ a_{21} \end{Bmatrix} \qquad (5.18)$$

where the second subscript refers to the mode. It is often convenient to let

$$a_{21} = 1 \qquad (5.19)$$

161

and the value of a_{11}, the modal component of M_1 in the first mode, may be found from the first of Equations 5.13. It reads

$$[(K_1 + K_2) - \omega_1^2 M_1] a_{11} - K_2 a_{21} = 0 \qquad (5.20)$$

Hence

$$a_{11} = \frac{K_2}{(K_1 + K_2) - \omega_1^2 M_1} a_{21}$$

or

$$a_{11} = \frac{K_2}{(K_1 + K_2) - \omega_1^2 M_1} \cdot 1 \qquad (5.21)$$

The second of Equations 5.13 reads

$$K_2 a_{11} = (K_2 - \omega_1^2 M_2) a_{21} \qquad (5.22)$$

and substitution of Equation 5.21 in Equation 5.22 gives rise to the original quadratic in ω^2, namely Equation 5.14, from which the natural frequency was found in the first place. In numerical calculations this step provides a useful check.

Similarly, the second mode is

$$\{a_2\} = \begin{Bmatrix} a_{12} \\ a_{22} \end{Bmatrix} \qquad (5.23)$$

and we may again, for convenience, assign

$$a_{22} = 1$$

So from Equations 5.13 we have

$$[(K_1 + K_2) - \omega_2^2 M_1] a_{12} - K_2 a_{22} = 0 \qquad (5.24)$$

or

$$a_{12} = \frac{K_2}{(K_1 + K_2) - \omega_2^2 M_1} \cdot 1 \qquad (5.25)$$

Example 5.2

Consider, again, the model shown in Figure 5.3a in which $K_1 = 3K$, $K_2 = 2K$ and $M_1 = M_2 = M$. From Equation 5.16 the natural frequency of the first mode is

$$\omega_1^2 = \frac{2MK + 5MK - \sqrt{[(2MK + 5MK)^2 - 24M^2K^2]}}{2M^2}$$

$$= \frac{7 - \sqrt{(49 - 24)}}{2} \frac{K}{M}$$

or

$$\omega_1 = \sqrt{\frac{K}{M}}$$

The associated mode shape is

$$\{a_1\} = \begin{Bmatrix} a_{11} \\ a_{21} \end{Bmatrix}$$

Letting $a_{21} = 1$, we have, from Equation 5.21

$$a_{11} = \frac{2K}{5K - \omega_1^2 M} \cdot 1$$

$$= \frac{2K}{5K - K}$$

$$= 1/2$$

Thus

$$\{a_1\} = \begin{Bmatrix} \frac{1}{2} \\ 1 \end{Bmatrix}$$

The natural frequency of the second mode is (see Eqn 5.17)

$$\omega_2^2 = \frac{7 + 5}{2} \frac{K}{M}$$

or

$$\omega_2 = \sqrt{6} \sqrt{\frac{K}{M}}$$

and the associated mode shape is

$$a_{22} = 1$$

$$a_{12} = \frac{2}{5 - 6} \cdot 1$$

$$= -2$$

Hence

$$\{a_2\} = \begin{Bmatrix} -2 \\ 1 \end{Bmatrix}$$

The results are shown in Figure 5.4.

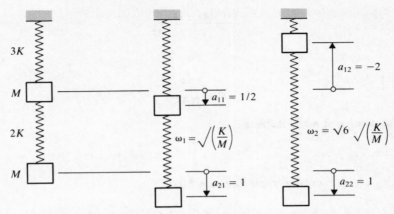

Figure 5.4 Natural frequencies and associated mode shapes.

5.4 Orthogonality of the mode shapes

Certain relationships exist between the modes of vibration which are exceedingly useful in uncoupling the equations of motion. Recall that, in writing the equations of motion for our simple 2 DOF system, coupling could come about either through the stiffness matrix, or through the mass matrix, it having been pointed out that, in general, both matrices could have off-diagonal terms. The equations of motion are therefore coupled, so that transformations which permit uncoupling greatly facilitate the solution of these equations.

Loosely speaking orthogonality is tantamount to saying that the scalar product of two vectors, which in this case are the vectors describing two different mode shapes, is zero. In fact, this scalar product is calculated with respect to a weighting supplied by either the mass matrix or the stiffness matrix. The details are as follows.

The equation for the first mode is

$$-\omega_1^2\begin{bmatrix} M_1 & 0 \\ 0 & M_2 \end{bmatrix}\begin{Bmatrix} a_{11} \\ a_{21} \end{Bmatrix} + \begin{bmatrix} (K_1+K_2) & -K_2 \\ -K_2 & K_2 \end{bmatrix}\begin{Bmatrix} a_{11} \\ a_{21} \end{Bmatrix} = \begin{Bmatrix} 0 \\ 0 \end{Bmatrix}$$

or (5.26)

$$-\omega_1^2[M]\{a_1\} + [K]\{a_1\} = 0$$

Similarly, the equation for the second mode is

$$-\omega_2^2[M]\{a_2\} + [K]\{a_2\} = 0 \qquad (5.27)$$

The transpose of the first-mode column vector is a row vector

$$\{a_1\}^T = \{a_{11} \quad a_{21}\}$$

Similarly,

$$\{a_2\}^T = \{a_{12} \quad a_{22}\}$$

Now premultiply Equation 5.26 by $\{a_2\}^T$, yielding

$$-\omega_1^2\{a_2\}^T[M]\{a_1\} + \{a_2\}^T[K]\{a_1\} = 0 \qquad (5.28)$$

Similarly, premultiply Equation 5.27 by $\{a_1\}^T$ to give

$$-\omega_2^2\{a_1\}^T[M]\{a_2\} + \{a_1\}^T[K]\{a_2\} = 0 \qquad (5.29)$$

Now transpose Equation 5.29. Term by term this is

$$\{-\omega_2^2\{a_1\}^T[M]\{a_2\}\}^T = -\omega_2^2\{a_2\}^T[M]^T\{\{a_1\}^T\}^T \qquad (5.30)$$

However, the mass matrix is symmetric so the interchanging of rows and columns leaves that matrix unchanged, and the transpose of a transposed vector yields the original vector. Thus, the right-hand side of Equation 5.30 is

$$-\omega_2^2\{a_2\}^T[M]\{a_1\}$$

Likewise the second entry on the left of Equation 5.29 when transposed is

$$\{a_2\}^T[K]\{a_1\}$$

Hence the transpose of Equation 5.29 is

$$-\omega_2^2\{a_2\}^T[M]\{a_1\} + \{a_2\}^T[K]\{a_1\} = 0 \qquad (5.31)$$

Subtracting Equation 5.31 from Equation 5.28 gives

$$(\omega_2^2 - \omega_1^2)\{a_2\}^T[M]\{a_1\} = 0 \qquad (5.32)$$

Now, almost invariably, $\omega_2 \neq \omega_1$, so we have

$$\{a_2\}^T[M]\{a_1\} = 0 \qquad (5.33)$$

and, if this is the case, then it follows from Equation 5.31 that

$$\{a_2\}^T[K]\{a_1\} = 0 \qquad (5.34)$$

165

also. Equations 5.33 and 5.34 are the orthogonality relations. In fact our proof is perfectly general and would hold for any multi-degree-of-freedom system: we need only replace the subscripts 1 and 2 by i and j and so obtain

$$\{a_j\}^T[M]\{a_i\} = 0$$

$$i \neq j \tag{5.35}$$

$$\{a_j\}^T[K]\{a_i\} = 0$$

Accordingly, if we form a matrix $[A]$ in which each column is a mode shape, that is if

$$[A] = [\{a_1\}\{a_2\} \ldots \{a_i\}\{a_j\} \ldots \{a_n\}]$$

then, for this n-degree-of-freedom system the orthogonality conditions can be written as

$$[A]^T[M][A] = [\backslash], \qquad [A]^T[K][A] = [\backslash] \tag{5.36}$$

where each matrix on the right-hand side is diagonal. In fact, by scaling the mode shapes it is possible to make the entries along the diagonal all unity and this is often done. That is, by suitable scaling of the mode shapes we can make

$$[A]^T[M][A] = [I] \tag{5.37}$$

where

$$[I] = \begin{bmatrix} 1 & & & & & \\ & 1 & & & & \\ & & 1 & & & \\ & & & \cdot & & \\ & & & & \cdot & \\ & & & & & \cdot \end{bmatrix}$$

The following example illustrates these points.

Example 5.3

Following on from Example 5.2, we have

$$[K] = \begin{bmatrix} 5K & -2K \\ -2K & 2K \end{bmatrix}, \qquad [M] = \begin{bmatrix} M & 0 \\ 0 & M \end{bmatrix}$$

166

or, ignoring dimensions,

$$[K] = \begin{bmatrix} 5 & -2 \\ -2 & 2 \end{bmatrix}, \qquad [M] = \begin{bmatrix} 1 & 0 \\ 0 & 1 \end{bmatrix}$$

The natural frequencies are

$$\omega_1 = \sqrt{\frac{K}{M}}, \qquad \omega_2 = \sqrt{6}\sqrt{\frac{K}{M}}$$

so $\omega_1 \neq \omega_2$.

The associated modes are

$$\{a_1\} = \begin{Bmatrix} \frac{1}{2} \\ 1 \end{Bmatrix}, \qquad \{a_2\} = \begin{Bmatrix} -2 \\ 1 \end{Bmatrix}$$

From Equation 5.33 the orthogonality relationship is

$$\{a_2\}^T [M] \{a_1\} = \{-2 \quad 1\} \begin{bmatrix} 1 & 0 \\ 0 & 1 \end{bmatrix} \begin{Bmatrix} \frac{1}{2} \\ 1 \end{Bmatrix}$$

$$= \{-2 \quad 1\} \begin{Bmatrix} \frac{1}{2} \\ 1 \end{Bmatrix}$$

$$= -1 + 1$$

$$= 0$$

Similarly,

$$\{a_2\}^T [K] \{a_1\} = \{-2 \quad 1\} \begin{bmatrix} 5 & -2 \\ -2 & 2 \end{bmatrix} \begin{Bmatrix} \frac{1}{2} \\ 1 \end{Bmatrix}$$

$$= \{-2 \quad 1\} \begin{Bmatrix} \frac{1}{2} \\ 1 \end{Bmatrix}$$

$$= -1 + 1$$

$$= 0$$

The modes of vibration are orthogonal.

From Equation 5.36.

$$[A]^T [M] [A] = \begin{bmatrix} \frac{1}{2} & 1 \\ -2 & 1 \end{bmatrix} \begin{bmatrix} 1 & 0 \\ 0 & 1 \end{bmatrix} \begin{bmatrix} \frac{1}{2} & -2 \\ 1 & 1 \end{bmatrix}$$

$$= \begin{bmatrix} \frac{1}{2} & 1 \\ -2 & 1 \end{bmatrix} \begin{bmatrix} \frac{1}{2} & -2 \\ 1 & 1 \end{bmatrix}$$

$$= \begin{bmatrix} \frac{5}{4} & 0 \\ 0 & 5 \end{bmatrix}$$

which is a diagonal. If we want this diagonal matrix to be the unit matrix, we can scale each mode shape. That is let

$$\{a_1\} = c_1 \begin{Bmatrix} \frac{1}{2} \\ 1 \end{Bmatrix}, \qquad \{a_2\} = c_2 \begin{Bmatrix} -2 \\ 1 \end{Bmatrix}$$

where c_1 and c_2 are constants to be determined. We want

$$\{a_1\}^T [M] \{a_1\} = 1$$

or

$$c_1^2 \{ \tfrac{1}{2} \quad 1 \} \begin{bmatrix} 1 & 0 \\ 0 & 1 \end{bmatrix} \begin{Bmatrix} \frac{1}{2} \\ 1 \end{Bmatrix} = 1$$

$$c_1^2 \{ \tfrac{1}{2} \quad 1 \} \begin{Bmatrix} \frac{1}{2} \\ 1 \end{Bmatrix} = 1$$

$$c_1^2 \times \tfrac{5}{4} = 1$$

$$c_1 = 2/\sqrt{5}$$

and

$$\{a_2\}^T [M] \{a_2\} = 1$$

or

$$c_2^2 \{ -2 \quad 1 \} \begin{bmatrix} 1 & 0 \\ 0 & 1 \end{bmatrix} \begin{Bmatrix} -2 \\ 1 \end{Bmatrix} = 1$$

$$c_2^2 \times 5 = 1$$

$$c_2 = 1/\sqrt{5}$$

Hence, the normalised mode shapes are

$$\{a_1\} = \frac{1}{\sqrt{5}} \begin{Bmatrix} 1 \\ 2 \end{Bmatrix}, \qquad \{a_2\} = \frac{1}{\sqrt{5}} \begin{Bmatrix} -2 \\ 1 \end{Bmatrix}$$

These two vectors are perpendicular (i.e. orthogonal) as may be seen from Figure 5.5.

5.5 Free vibrational response

In matrix form the equations of motion for free vibrations are

$$[M] \{\ddot{\Delta}\} + [K] \{\Delta\} = \{0\} \tag{5.38}$$

and the solution of these equations is sought for given initial conditions. We are talking here, generally, of an n-degree-of-freedom system so the initial conditions for displacement consist of one value for each initial displacement, which is n values in all. So, too, we require n initial velocities, one per degree of freedom. That is

$$\{\Delta(0)\} = \begin{Bmatrix} \Delta_1(0) \\ \Delta_2(0) \\ \Delta_3(0) \\ \vdots \\ \Delta_n(0) \end{Bmatrix} \tag{5.39}$$

and

$$\{\dot{\Delta}(0)\} = \begin{Bmatrix} \dot{\Delta}_1(0) \\ \dot{\Delta}_2(0) \\ \dot{\Delta}_3(0) \\ \vdots \\ \dot{\Delta}_n(0) \end{Bmatrix} \tag{5.40}$$

The equations of motion are coupled and it is not possible to proceed to a formal solution (except, of course, by purely numerical means which lie outside the scope of this work – see Biggs (1964), Clough and Penzien (1975), Craig (1981) and Paz (1980), for example) unless we uncouple the equations. The orthogonality relationships developed in the previous section are here most useful. Let us define a new set of displacements $\{\xi\}$, where

$$\{\Delta\} = [A]\{\xi\} \tag{5.41}$$

and $[A]$ is the matrix of n columns in which each column is a mode shape of the system. Substituting this in Equation 5.38 yields

$$[M][A]\{\ddot{\xi}\} + [K][A]\{\xi\} = \{0\} \tag{5.42}$$

Now premultiply Equation 5.42 by the transpose of A,

$$[A]^T[M][A]\{\ddot{\xi}\} + [A]^T[K][A]\{\xi\} = \{0\} \tag{5.43}$$

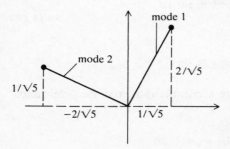

Figure 5.5 Orthogonality of mode shapes.

Now, from our orthogonality relationships, both $[A]^T[M][A]$ and $[A]^T[K][A]$ are diagonal matrices, and we have uncoupled the equations of motion. They read

$$\{a_1\}^T[M]\{a_1\}\ddot{\xi}_1 + \{a_1\}^T[K]\{a_1\}\xi_1 = 0$$

$$\{a_2\}^T[M]\{a_2\}\ddot{\xi}_2 + \{a_2\}^T[K]\{a_2\}\xi_2 = 0$$

$$\vdots \qquad\qquad \vdots$$

$$\{a_n\}^T[M]\{a_n\}\ddot{\xi}_n + \{a_n\}^T[K]\{a_n\}\xi_n = 0$$

(5.44)

or

$$\ddot{\xi}_1 + \frac{\{a_1\}^T[K]\{a_1\}}{\{a_1\}^T[M]\{a_1\}}\xi_1 = 0,$$

(5.45)

etc. The quantity

$$\frac{\{a_1\}^T[K]\{a_1\}}{\{a_1\}^T[M]\{a_1\}}$$

is called Rayleigh's quotient and is equal to ω_1^2, the square of the natural frquency of the first mode. Accordingly, our equations are:

$$\ddot{\xi}_1 + \omega_1^2\xi_1 = 0$$

$$\ddot{\xi}_2 + \omega_2^2\xi_2 = 0$$

$$\vdots$$

$$\ddot{\xi}_n + \omega_n^2\xi_n = 0$$

(5.46)

The solution of Equations 5.46 is

$$\xi_1 = A_1\cos\omega_1 t + B_1\sin\omega_1 t$$

$$\xi_2 = A_2\cos\omega_2 t + B_2\sin\omega_2 t$$

$$\vdots$$

$$\xi_n = A_n\cos\omega_n t + B_n\sin\omega_n t$$

(5.47)

where A_1, B_1, A_2, B_2, ..., A_n, B_n are constants that may be determined from the initial conditions. We have

$$\{\Delta(0)\} = [A]\{\xi(0)\}$$

which may be premultiplied by $[A]^T[M]$ giving

$$[A]^T[M][A]\{\xi(0)\} = [A]^T[M]\{\Delta(0)\}$$

If we normalise the mode shapes such that

$$[A]^T[M][A] = [I],$$

and this was suggested in the previous section, then

$$\{\xi(0)\} = [A]^T[M]\{\Delta(0)\} \tag{5.48}$$

and, similarly

$$\{\dot{\xi}(0)\} = [A]^T[M]\{\dot{\Delta}(0)\} \tag{5.49}$$

Hence, because $\{\Delta(0)\}$ and $\{\dot{\Delta}(0)\}$ are prescribed, we can find $\{\xi(0)\}$ and $\{\dot{\xi}(0)\}$, and so evaluate the constants of integration. We then find the actual displacements from $\{\Delta\} = [A]\{\xi\}$ because $\{\xi\}$ is now known.

Example 5.4

The equations of motion of our simple model are

$$\begin{bmatrix} M & 0 \\ 0 & M \end{bmatrix} \begin{Bmatrix} \ddot{\Delta}_1 \\ \ddot{\Delta}_2 \end{Bmatrix} + \begin{bmatrix} 5K & -2K \\ -2K & 2K \end{bmatrix} \begin{Bmatrix} \Delta_1 \\ \Delta_2 \end{Bmatrix} = \begin{Bmatrix} 0 \\ 0 \end{Bmatrix}$$

We need a solution to these equations for the free vibrations that ensue from initial conditions which, for this example, are prescribed as

$$\begin{Bmatrix} \Delta_1(0) \\ \Delta_2(0) \end{Bmatrix} = \begin{Bmatrix} 0 \\ \Delta \end{Bmatrix}, \qquad \begin{Bmatrix} \dot{\Delta}_1(0) \\ \dot{\Delta}_2(0) \end{Bmatrix} = \begin{Bmatrix} 0 \\ 0 \end{Bmatrix}$$

Recall that the natural frequencies and associated normalised mode shapes are

$$\omega_1 = \sqrt{\left(\frac{K}{M}\right)}, \qquad \begin{Bmatrix} a_{11} \\ a_{21} \end{Bmatrix} = \frac{1}{\sqrt{5}} \begin{Bmatrix} 1 \\ 2 \end{Bmatrix}$$

$$\omega_2 = \sqrt{6}\sqrt{\left(\frac{K}{M}\right)}, \qquad \begin{Bmatrix} a_{12} \\ a_{22} \end{Bmatrix} = \frac{1}{\sqrt{5}} \begin{Bmatrix} -2 \\ 1 \end{Bmatrix}$$

Hence the matrix of mode shapes is

$$[A] = \frac{1}{\sqrt{5}} \begin{bmatrix} 1 & -2 \\ 2 & 1 \end{bmatrix}$$

and we now make the substitution

$$\{\Delta\} = [A]\{\xi\}$$

that is

$$\begin{Bmatrix} \Delta_1 \\ \Delta_2 \end{Bmatrix} = \frac{1}{\sqrt{5}} \begin{bmatrix} 1 & -2 \\ 2 & 1 \end{bmatrix} \begin{Bmatrix} \xi_1 \\ \xi_2 \end{Bmatrix}$$

yielding

$$\frac{1}{\sqrt{5}} \begin{bmatrix} M & 0 \\ 0 & M \end{bmatrix} \begin{bmatrix} 1 & -2 \\ 2 & 1 \end{bmatrix} \begin{Bmatrix} \ddot{\xi}_1 \\ \ddot{\xi}_2 \end{Bmatrix} + \frac{1}{\sqrt{5}} \begin{bmatrix} 5K & -2K \\ -2K & 2K \end{bmatrix} \begin{bmatrix} 1 & -2 \\ 2 & 1 \end{bmatrix} \begin{Bmatrix} \xi_1 \\ \xi_2 \end{Bmatrix} = \begin{Bmatrix} 0 \\ 0 \end{Bmatrix}$$

Now premultiply by

$$[A]^T = \frac{1}{\sqrt{5}} \begin{bmatrix} 1 & 2 \\ -2 & 1 \end{bmatrix}$$

giving

$$\frac{1}{5} \begin{bmatrix} 1 & 2 \\ -2 & 1 \end{bmatrix} \begin{bmatrix} M & 0 \\ 0 & M \end{bmatrix} \begin{bmatrix} 1 & -2 \\ 2 & 1 \end{bmatrix} \begin{Bmatrix} \ddot{\xi}_1 \\ \ddot{\xi}_2 \end{Bmatrix} + \frac{1}{5} \begin{bmatrix} 1 & 2 \\ -2 & 1 \end{bmatrix} \begin{bmatrix} 5K & -2K \\ -2K & 2K \end{bmatrix} \begin{bmatrix} 1 & -2 \\ 2 & 1 \end{bmatrix} \begin{Bmatrix} \xi_1 \\ \xi_2 \end{Bmatrix} = \begin{Bmatrix} 0 \\ 0 \end{Bmatrix}$$

That is

$$\frac{1}{5} \begin{bmatrix} 1 & 2 \\ -2 & 1 \end{bmatrix} \begin{bmatrix} M & -2M \\ 2M & M \end{bmatrix} \begin{Bmatrix} \ddot{\xi}_1 \\ \ddot{\xi}_2 \end{Bmatrix} + \frac{1}{5} \begin{bmatrix} 1 & 2 \\ -2 & 1 \end{bmatrix} \begin{bmatrix} K & -12K \\ 2K & 6K \end{bmatrix} \begin{Bmatrix} \xi_1 \\ \xi_2 \end{Bmatrix} = \begin{Bmatrix} 0 \\ 0 \end{Bmatrix}$$

or

$$\frac{1}{5} \begin{bmatrix} 5M & 0 \\ 0 & 5M \end{bmatrix} \begin{Bmatrix} \ddot{\xi}_1 \\ \ddot{\xi}_2 \end{Bmatrix} + \frac{1}{5} \begin{bmatrix} 5K & 0 \\ 0 & 30K \end{bmatrix} \begin{Bmatrix} \xi_1 \\ \xi_2 \end{Bmatrix} = \begin{Bmatrix} 0 \\ 0 \end{Bmatrix}$$

or

$$M\ddot{\xi}_1 + K\xi_1 = 0$$

and

$$M\ddot{\xi}_2 + 6K\xi_2 = 0$$

That is,

$$\ddot{\xi}_1 + \omega_1^2 \xi_1 = 0, \qquad \ddot{\xi}_2 + \omega_2^2 \xi_2 = 0$$

or

$$\ddot{\xi}_1 + \omega^2 \xi_1 = 0, \qquad \ddot{\xi}_2 + 6\omega^2 \xi_2 = 0$$

where $\omega^2 = K/M$. The solutions are

$$\xi_1 = A_1 \cos \omega t + B_1 \sin \omega t$$

$$\xi_2 = A_2 \cos \sqrt{6}\omega t + B_2 \sin \sqrt{6}\omega t$$

The initial conditions are

$$\begin{Bmatrix} \xi_1(0) \\ \xi_2(0) \end{Bmatrix} = [A]^T [M] \begin{Bmatrix} 0 \\ \Delta \end{Bmatrix} = \frac{1}{\sqrt{5}} \begin{bmatrix} 1 & 2 \\ -2 & 1 \end{bmatrix} \begin{bmatrix} 1 & 0 \\ 0 & 1 \end{bmatrix} \begin{Bmatrix} 0 \\ \Delta \end{Bmatrix}$$

$$= \frac{1}{\sqrt{5}} \begin{bmatrix} 1 & 2 \\ -2 & 1 \end{bmatrix} \begin{Bmatrix} 0 \\ \Delta \end{Bmatrix}$$

$$= \frac{1}{\sqrt{5}} \begin{Bmatrix} 2\Delta \\ \Delta \end{Bmatrix}$$

and

$$\begin{Bmatrix} \dot{\xi}_1(0) \\ \dot{\xi}_2(0) \end{Bmatrix} = [A]^T [M] \begin{Bmatrix} 0 \\ 0 \end{Bmatrix} = \begin{Bmatrix} 0 \\ 0 \end{Bmatrix}$$

Hence

$$A_1 = 2\Delta/\sqrt{5} \qquad A_2 = \Delta/\sqrt{5}$$

$$B_1 = 0, \qquad B_2 = 0$$

and the solution to the uncoupled equations is

$$\xi_1 = \frac{2\Delta}{\sqrt{5}} \cos \omega t, \qquad \xi_2 = \frac{\Delta}{\sqrt{5}} \cos \sqrt{6}\,\omega t$$

In terms of the physical displacements we have

$$\begin{Bmatrix} \Delta_1 \\ \Delta_2 \end{Bmatrix} = [A] \begin{Bmatrix} \xi_1 \\ \xi_2 \end{Bmatrix} = \frac{1}{\sqrt{5}} \begin{bmatrix} 1 & -2 \\ 2 & 1 \end{bmatrix} \begin{Bmatrix} \xi_1 \\ \xi_2 \end{Bmatrix}$$

$$= \frac{1}{\sqrt{5}} \begin{Bmatrix} \xi_1 - 2\xi_2 \\ 2\xi_1 + \xi_2 \end{Bmatrix} = \frac{1}{5} \begin{Bmatrix} 2\Delta \cos \omega t - 2\Delta \cos \sqrt{6}\,\omega t \\ 4\Delta \cos \omega t + \Delta \cos \sqrt{6}\,\omega t \end{Bmatrix}$$

or

$$\Delta_1(t) = \frac{2\Delta}{5} (\cos \omega t - \cos \sqrt{6}\omega t)$$

$$\Delta_2(t) = \frac{\Delta}{5} (4 \cos \omega t + \cos \sqrt{6}\,\omega t)$$

173

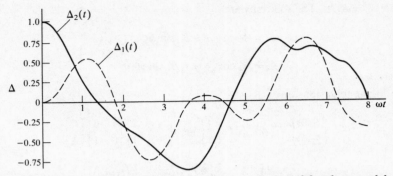

Figure 5.6 Free vibrational response of a two-degree-of-freedom model.

Notice that the initial conditions are satisfied, as expected. Notice, further, with respect to Figure 5.6 that the time histories of displacement for each mass have no obvious periodicity, even though they are made up of modal components with clearly defined periodicity.

5.6 Rayleigh's quotient and the energy of vibration

We saw in the previous section that the transformation from the physical displacements $\{\Delta\}$ to another set $\{\xi\}$ according to the relationships

$$\{\Delta\} = [A]\{\xi\}$$

led to the uncoupling of the equations of motion. At this stage, however, it is difficult to ascribe any particular meaning to the displacements $\{\xi\}$. However, consider the energy of vibration.

The kinetic energy is

$$\text{KE} = \tfrac{1}{2}\{\dot{\Delta}\}^{\text{T}}[M]\{\dot{\Delta}\} \qquad (5.50)$$

As an aside, consider the example with which we started this chapter, where

$$\text{KE} = \tfrac{1}{2}M_1\dot{\Delta}_1^2 + \tfrac{1}{2}M_2\dot{\Delta}_2^2$$

From Equation 5.50 we have

$$\text{KE} = \tfrac{1}{2}\{\dot{\Delta}_1 \quad \dot{\Delta}_2\}\begin{bmatrix} M_1 & 0 \\ 0 & M_2 \end{bmatrix}\begin{Bmatrix} \dot{\Delta}_1 \\ \dot{\Delta}_2 \end{Bmatrix}$$

174

$$= \tfrac{1}{2}\{\dot{\Delta}_1 \quad \dot{\Delta}_2\}\begin{Bmatrix} M_1\dot{\Delta}_1 \\ M_2\dot{\Delta}_2 \end{Bmatrix}$$

$$= \tfrac{1}{2}M_1\dot{\Delta}_1^2 + \tfrac{1}{2}M_2\dot{\Delta}_2^2,$$

as expected. Now apply the co-ordinate transformation

$$\{\Delta\} = [A]\{\xi\}$$

to Equation 5.50 noting that

$$\{\dot{\Delta}\} = [A]\{\dot{\xi}\}$$

and

$$\{\dot{\Delta}\}^T = \{[A]\{\dot{\xi}\}\}^T$$

$$= \{\dot{\xi}\}[A]^T$$

Hence

$$KE = \tfrac{1}{2}\{\dot{\xi}\}^T[A]^T[M][A]\{\dot{\xi}\} \tag{5.51}$$

But $[A]^T[M][A]$ is a diagonal matrix and so the right-hand side of Equation 5.51 is sum of squares of each component in $\{\dot{\xi}\}$. Furthermore, if we had normalised the mode shapes such that

$$[A]^T[M][A] = [I]$$

we would have

$$KE = \tfrac{1}{2}\{\dot{\xi}\}^T[I]\{\dot{\xi}\}$$

or

$$KE = \tfrac{1}{2}(\dot{\xi}_1^2 + \dot{\xi}_2^2 + \ldots + \dot{\xi}_n^2) \tag{5.52}$$

In the case of the potential energy we have

$$PE = \tfrac{1}{2}\{\Delta\}^T[K]\{\Delta\}$$

Again, as an aside, consider our original example

$$PE = \tfrac{1}{2}K_1\Delta_1^2 + \tfrac{1}{2}K_2(\Delta_2 - \Delta_1)^2$$

$$= \tfrac{1}{2}K_1\Delta_1^2 + \tfrac{1}{2}K_2\Delta_2^2 - K_2\Delta_1\Delta_2 + \tfrac{1}{2}K_2\Delta_1^2$$

or

$$PE = \tfrac{1}{2}(K_1 + K_2)\Delta_1^2 - K_2\Delta_1\Delta_2 + \tfrac{1}{2}K_2\Delta_2^2$$

From the matrix expression we have

$$PE = \tfrac{1}{2}\{\Delta_1 \quad \Delta_2\}\begin{bmatrix}(K_1 + K_2) & -K_2 \\ -K_2 & K_2\end{bmatrix}\begin{Bmatrix}\Delta_1 \\ \Delta_2\end{Bmatrix}$$

$$= \tfrac{1}{2}\{\Delta_1 \quad \Delta_2\}\begin{Bmatrix}(K_1 + K_2)\Delta_1 & -K_2\Delta_2 \\ -K_2\Delta_1 & +K_2\Delta_2\end{Bmatrix}$$

$$= \tfrac{1}{2}[(K_1 + K_2)\Delta_1^2 - K_2\Delta_2\Delta_1 - K_2\Delta_1\Delta_2 + K_2\Delta_2^2]$$

$$= \tfrac{1}{2}(K_1 + K_2)\Delta_1^2 - K_2\Delta_1\Delta_2 + \tfrac{1}{2}K_2\Delta_2^2,$$

which is the same. Applying the transformation, we have

$$PE = \tfrac{1}{2}\{\xi\}^T[A]^T[K][A]\{\xi\} \tag{5.53}$$

and, again, on account of orthogonality, we know that $[A]^T[K][A]$ is a diagonal matrix. In fact, because we are assuming that the mode shapes have been normalised, the elements on the diagonal are the squares of the natural frequencies. That is

$$PE = \tfrac{1}{2}\{\xi\}^T\begin{bmatrix}\omega_1^2 & & & \\ & \omega_2^2 & & \\ & & \cdot & \\ & & & \cdot \\ & & & & \omega_n^2\end{bmatrix}\{\xi\}$$

or

$$PE = \tfrac{1}{2}(\omega_1^2\xi_1^2 + \omega_2^2\xi_2^2 + \ldots + \omega_n^2\xi_n^2) \tag{5.54}$$

Thus, the new displacement quantities allow the expressions for energy of vibration to be written as sums of squares of each component. It is for this

176

reason that the set of displacements is referred to as the 'normal co-ordinates' of the problem.

Note that if the mode shapes are not normalised according to the relationship $[A]^T[M][A] = [I]$, the natural frequency of the ith mode is given by

$$\omega_i^2 = \frac{\{a_i\}^T[K]\{a_i\}}{\{a_i\}^T[M]\{a_i\}}, \qquad i = 1, 2, \ldots, n \qquad (5.55)$$

This is Rayleigh's quotient. For illustrative purpose, and at the risk of going round in a circle, consider the first mode of our numerical problem. This we established as

$$\{a_1\} = \begin{Bmatrix} \frac{1}{2} \\ 1 \end{Bmatrix}$$

and from Equation 5.55, we have

$$\omega_1^2 = \frac{\{\frac{1}{2} \; 1\} \begin{bmatrix} 5K & -2K \\ -2K & 2K \end{bmatrix} \begin{Bmatrix} \frac{1}{2} \\ 1 \end{Bmatrix}}{\{\frac{1}{2} \; 1\} \begin{bmatrix} M & 0 \\ 0 & M \end{bmatrix} \begin{Bmatrix} \frac{1}{2} \\ 1 \end{Bmatrix}}$$

$$= \{\tfrac{1}{2} \; 1\} \begin{Bmatrix} K/2 \\ K \end{Bmatrix} \bigg/ \{\tfrac{1}{2} \; 1\} \begin{Bmatrix} M/2 \\ M \end{Bmatrix}$$

or

$$\omega_1^2 = \frac{5K/4}{5M/4} = \frac{K}{M}$$

which is the correct answer.

Rayleigh's quotient allows the generalisation of the approach outlined in Section 5.2 (Eqn 5.12) where an approximate method for determining the natural frequency of the fundamental mode of a 2 DOF system was presented. Suppose an estimate exists for the fundamental mode of a multi-degree-of-freedom system. Then, according to Rayleigh's theorem,

$$\omega_1^2 \leq \frac{\{a_1\}^T[K]\{a_1\}}{\{a_1\}^T[M]\{a_1\}}$$

where $\{a_1\}$ is the estimate of the first mode shape.

177

Example 5.5

Figure 5.7 shows a 3 DOF system. The equations of motion for free vibration are

$$\begin{bmatrix} M & 0 & 0 \\ 0 & M & 0 \\ 0 & 0 & M \end{bmatrix} \begin{Bmatrix} \ddot{\Delta}_1 \\ \ddot{\Delta}_2 \\ \ddot{\Delta}_3 \end{Bmatrix} \begin{bmatrix} 2K & -K & 0 \\ -K & 2K & -K \\ 0 & -K & K \end{bmatrix} \begin{Bmatrix} \Delta_1 \\ \Delta_2 \\ \Delta_3 \end{Bmatrix} = \begin{Bmatrix} 0 \end{Bmatrix}$$

An estimate for the first mode may be found by calculating the deflections due to gravity. That is we solve the equations

$$\begin{bmatrix} 2K & -K & 0 \\ -K & 2K & -K \\ 0 & -K & K \end{bmatrix} \begin{Bmatrix} a_{11} \\ a_{21} \\ a_{31} \end{Bmatrix} = \begin{bmatrix} M & 0 & 0 \\ 0 & M & 0 \\ 0 & 0 & M \end{bmatrix} \begin{Bmatrix} g \\ g \\ g \end{Bmatrix} = \begin{Bmatrix} Mg \\ Mg \\ Mg \end{Bmatrix}$$

This yields deflections that are in the ratio, $1:\frac{5}{3}:2$. So, a first estimate of the mode is

$$\{a_1\}_1 = \begin{Bmatrix} 1 \\ \frac{5}{3} \\ 2 \end{Bmatrix}$$

Rather than calculate the corresponding estimate of ω_1 via Rayleigh's quotient, we can miss this step and obtain a second estimate of the mode shape. Recall that, for the first mode, we have

$$[K]\{a_1\} = \omega_1^2 [M]\{a_1\}$$

Now ω_1^2 is a constant and its value need not be known at this stage. We can solve for the second estimate of $\{a_1\}$, namely $\{a_1\}_2$, by solving the equations

$$[K]\{a_1\}_2 = \omega_1^2 [M]\{a_1\}_1$$

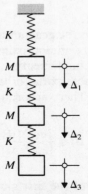

Figure 5.7 Definition diagram for a 3 DOF system.

This yields deflections that are in the ratio $14:25:31$ or

$$\{a_1\}_2 = \begin{Bmatrix} 1 \\ 1.79 \\ 2.21 \end{Bmatrix}$$

From Rayleigh's quotient

$$\omega_1^2 \leqslant \frac{\{1 \quad 1.79 \quad 2.21\} \begin{bmatrix} 2 & -1 & 0 \\ -1 & 2 & -1 \\ 0 & -1 & 1 \end{bmatrix} \begin{Bmatrix} 1 \\ 1.79 \\ 2.21 \end{Bmatrix}}{\{1 \quad 1.79 \quad 2.21\} \begin{bmatrix} 1 & 0 & 0 \\ 0 & 1 & 0 \\ 0 & 0 & 1 \end{bmatrix} \begin{Bmatrix} 1 \\ 1.79 \\ 2.21 \end{Bmatrix}} \cdot \frac{K}{M}$$

$$\omega_1^2 \leqslant \frac{\{1 \quad 1.79 \quad 2.21\} \begin{Bmatrix} 0.21 \\ 0.37 \\ 0.42 \end{Bmatrix}}{\{1 \quad 1.79 \quad 2.21\} \begin{Bmatrix} 1 \\ 1.79 \\ 2.21 \end{Bmatrix}} \cdot \frac{K}{M}$$

$$\omega_1^2 \leqslant \frac{1.80K}{9.09M} \leqslant 0.198 \frac{K}{M}$$

$$\omega_1 \leqslant 0.445 \sqrt{\left(\frac{K}{M}\right)}$$

This is very close to the correct answer of $\omega_1 = 0.45\sqrt{(K/M)}$ and

$$\{a_1\} = \begin{Bmatrix} 1 \\ 1.80 \\ 2.24 \end{Bmatrix}$$

5.7 Forced response of a multi-degree-of-freedom system

The equations of motion when forcing is present has the form

$$[M]\{\ddot{\Delta}\} + [K]\{\Delta\} = \{F\}$$

where $\{F\}$ is a vector of forces, one per degree of freedom, which are functions of time. We ignore damping for the present: it will be included later. We shall assume that we have zero initial conditions, i.e. $\{\dot{\Delta}\} = \{\Delta\} = \{0\}$.

179

If this is not the case we need only include the free vibrational response, as outlined previously, via the principle of superposition.

The equations of motion are coupled, and we uncouple them using the transformation to normal co-ordinates. This yields

$$[A]^T[M][A]\{\ddot{\xi}\} + [A]^T[K][A]\{\xi\} = [A]^T\{F\} \tag{5.56}$$

or

$$\ddot{\xi}_1 + \omega_1^2\xi_1 = \frac{\{a_1\}^T\{F\}}{\{a_1\}^T[M]\{a_1\}}$$

$$\ddot{\xi}_2 + \omega_2^2\xi_2 = \frac{\{a_2\}^T\{F\}}{\{a_2\}^T[M]\{a_2\}}$$

$$\vdots \qquad \vdots \qquad\qquad \vdots \tag{5.57}$$

$$\ddot{\xi}_n + \omega_n^2\xi_n = \frac{\{a_n\}^T\{F\}}{\{a_n\}^T[M]\{a_n\}}$$

Damping is most easily included at this point simply by including modal damping terms of the form $2\zeta_i\omega_i\dot{\xi}_i$ where ζ_i is the fraction of critical damping in the ith mode. At best an educated guess can be made as to the value of ζ_i: note, however, that simple assumptions like ζ_i = constant (for all modes) will tend to wipe out response in the higher modes which may, or may not, square with the facts.

Thus the equation of motion for the ith normal co-ordinate is

$$\ddot{\xi}_i + 2\zeta_i\omega_i\dot{\xi}_i + \omega_i^2\xi_i = \alpha_i, \qquad i = 1, 2, \ldots, n \tag{5.58}$$

where

$$\alpha_i = \{a_i\}^T\{F\}/\{a_i\}^T[M]\{a_i\},$$

which indicates the extent to which forcing is present in the ith mode. The general solution subject to $\xi_i(0) = \dot{\xi}_i(0) = 0$ is, from Duhamel's integral,

$$\xi_i(t) = \frac{1}{\omega_{d,i}} \int_0^t \alpha_i(\tau) \exp[-\zeta_i\omega_i(t-\tau)] \sin[\omega_{d,i}(t-\tau)] \, d\tau \tag{5.59}$$

where $\omega_{d,i} = \omega_i \sqrt{(1-\zeta_i^2)}$. Finally we form

$$\{\Delta\} = [A]\{\xi\}$$

to obtain time histories of response for each modal displacement.

Since there is not much more that needs to be said, it is best to move directly to an example.

Example 5.6

Suppose two constant forces, F, are suddenly applied to the spring–mass system shown in Figure 5.8. Ignoring damping the equations of motion are

$$\begin{bmatrix} M & 0 \\ 0 & M \end{bmatrix} \begin{Bmatrix} \ddot{\Delta}_1 \\ \ddot{\Delta}_2 \end{Bmatrix} + \begin{bmatrix} 5K & -2K \\ -2K & 2K \end{bmatrix} \begin{Bmatrix} \Delta_1 \\ \Delta_2 \end{Bmatrix} = \begin{Bmatrix} F \\ F \end{Bmatrix}$$

The static deflections are

$$\Delta_{1,st} = 4F/6K$$

$$\Delta_{2,st} = 7F/6K$$

Now, from our earlier work,

$$\omega_1^2 = K/M, \qquad \{a_1\} = \frac{1}{\sqrt{5}} \begin{Bmatrix} 1 \\ 2 \end{Bmatrix}$$

$$\omega_2^2 = 6K/M, \qquad \{a_2\} = \frac{1}{\sqrt{5}} \begin{Bmatrix} -2 \\ 1 \end{Bmatrix}$$

Hence, applying the transformation

$$\{\Delta\} = [A]\{\xi\}$$

yields

$$\ddot{\xi}_1 + \omega^2 \xi_1 = \alpha_1, \qquad \ddot{\xi}_2 + 6\omega^2 \xi_2 = \alpha_2$$

where $\omega^2 = K/M$, and

$$\alpha_1 = \frac{\{a_1\}^T \{F\}}{\{a_1\}^T [M] \{a_1\}} = \frac{\frac{1}{\sqrt{5}} \{1 \quad 2\} \begin{Bmatrix} F \\ F \end{Bmatrix}}{\frac{1}{5} \{1 \quad 2\} \begin{bmatrix} M & \\ & M \end{bmatrix} \begin{Bmatrix} 1 \\ 2 \end{Bmatrix}}$$

$$= 3F/\sqrt{5}M = \omega^2 3F/\sqrt{5}K$$

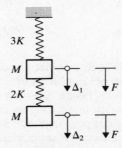

Figure 5.8

181

and

$$\alpha_2 = \frac{\{a_2\}^{\text{T}}\{F\}}{\{a_2\}^{\text{T}}[M]\{a_2\}} = \frac{\frac{1}{\sqrt{5}}\{-2 \quad 1\}\begin{Bmatrix} F \\ F \end{Bmatrix}}{M}$$

$$= -F/\sqrt{5}M = -6\omega^2(1/6\sqrt{5})(F/K)$$

The solutions, for zero initial conditions, are

$$\xi_1(t) = \frac{3}{\sqrt{5}}\frac{F}{K}(1 - \cos \omega t)$$

$$\xi_2(t) = -\frac{1}{6\sqrt{5}}\frac{F}{K}(1 - \cos \sqrt{6}\,\omega t)$$

Returning to the actual displacements we have

$$\begin{Bmatrix} \Delta_1(t) \\ \Delta_2(t) \end{Bmatrix} = \frac{1}{\sqrt{5}}\begin{bmatrix} 1 & -2 \\ 2 & 1 \end{bmatrix}\begin{Bmatrix} \xi_1(t) \\ \xi_2(t) \end{Bmatrix}$$

or

$$\Delta_1(t) = \frac{3}{5}\frac{F}{K}(1 - \cos \omega t) + \frac{1}{15}\frac{F}{K}(1 - \cos \sqrt{6}\omega t)$$

$$\Delta_2(t) = \frac{6}{5}\frac{F}{K}(1 - \cos \omega t) - \frac{1}{30}\frac{F}{K}(1 - \cos \sqrt{6}\omega t)$$

There are several points to note. First, the vibration will not have a clearly defined periodicity although the second-mode contributions are significantly smaller than those of the first mode. This may be seen from the ratios $\frac{1}{15}:\frac{3}{5}$ and $\frac{1}{30}:\frac{6}{5}$, respectively. This is often the case in practical vibration problems. The fundamental mode dominates the response. Secondly, with time, damping will wipe out the cosine terms and we are left with

$$\Delta_1 \rightarrow (\tfrac{3}{5} + \tfrac{1}{15})\,F/K \rightarrow 4F/6K$$

$$\Delta_2 \rightarrow (\tfrac{6}{5} - \tfrac{1}{30})\,F/K \rightarrow 7F/6K$$

which are the static solutions. This provides a good indication that we have done the analysis correctly. Another point, which is of little more than academic interest, is that the maximum displacements are not necessarily twice the static displacements: they can be slightly more, and they can be slightly less. Furthermore, in general, the peak value of one displacement will not occur simultaneously with the peak value of the other. This minor complication is due to the presence of the second mode.

182

Finally, the accelerations are

$$\ddot{\Delta}_1(t) = \frac{3}{5}\frac{F}{M}\cos \omega t + \frac{2}{5}\frac{F}{M}\cos \sqrt{6}\,\omega t$$

$$\ddot{\Delta}_2(t) = \frac{6}{5}\frac{F}{M}\cos \omega t - \frac{1}{5}\frac{F}{M}\cos \sqrt{6}\omega t$$

The second-mode contributions are, proportionally, a good deal more significant, which was a point made at the end of Example 1.3. At time $t = 0$

$$\ddot{\Delta}_1(0) = (\tfrac{3}{5} + \tfrac{2}{5})\,F/M = F/M$$

$$\ddot{\Delta}_2(0) = (\tfrac{6}{5} - \tfrac{1}{5})\,F/M = F/M$$

and this is as expected, since there can be no spring forces because the springs have not yet stretched.

Example 5.7 *Dynamic response of an offshore crane in hoisting mode*

A schematic of an offshore crane is shown in Figure 5.9. First we obtain the equations of motion for free vibration for this structure modelled as a 2 DOF system. Then for a given case we obtain the natural frequencies and mode shapes and move on to consider peak dynamic response generated during a particular hoisting operation. The example touches on, and helps clarify, various points made in this chapter.

EQUATIONS OF MOTION FOR FREE VIBRATION

We select as the two degrees of freedom the angular (rigid-body) rotation of the boom, θ, and the vertical deflection of the payload, Δ_2 (see Fig. 5.10). The vertical movement of the tip of the boom is

$$\Delta_1 = (l_3 \cos \alpha)\theta$$

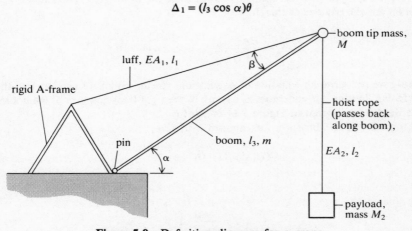

Figure 5.9 Definition diagram for a crane.

183

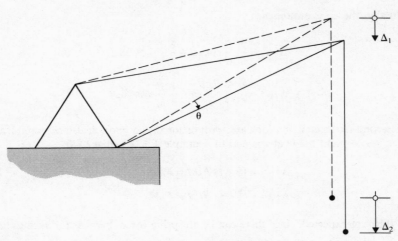

Figure 5.10 The 2 DOF model.

and, as we shall see later, this is a more convenient displacement quantity to work with once the equations of motion have been set up.

When the payload displaces vertically by Δ_2, and there is a corresponding vertical movement of the boom tip of $(l_3 \cos \alpha)\, \theta$, the net lengthening of the hoist rope is

$$\Delta_2 - (l_3 \cos \alpha)\theta$$

and the tensile force developed is

$$\frac{EA_2}{l_2 + l_3} \left[\Delta_2 - (l_3 \cos \alpha)\theta \right]$$

where EA_2 is an 'effective' axial stiffness based on the manner in which the pulley system is deployed (i.e. the number of falls from the boom tip). The equation of motion for the payload is therefore

$$M_2 \ddot{\Delta}_2 + \frac{EA_2}{l_2 + l_3} \left[\Delta_2 - (l_3 \cos \alpha)\theta \right] = 0$$

Before we can write an equation of equilibrium for the boom we need to find the stretching that the luff undergoes as a result of the rigid-body rotation of the boom. The definition diagram in Figure 5.11 is helpful.

Moment equilibrium for a unit applied load gives

$$T(l_3 \sin \beta) = (l_3 \cos \alpha) 1$$

or

$$T = \frac{\cos \alpha}{\sin \beta} \cdot 1$$

184

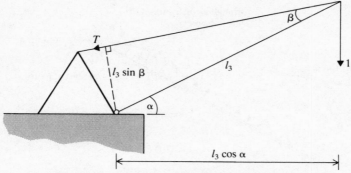

Figure 5.11   Definition diagram for luff tension.

Consider a virtual displacement which is a rigid-body rotation of the boom, say θ. The displacement of the unit load is $(l_3 \cos \alpha)\theta$ while the lengthening of the luff is Δl_1. From the virtual work theorem

$$T \cdot \Delta l_1 = 1 \cdot (l_3 \cos \alpha)\theta$$

or

$$\Delta l_1 = \frac{(l_3 \cos \alpha)\theta}{\cos \alpha / \sin \beta} = (l_3 \sin \beta)\theta$$

In retrospect, this could have been established from kinematics alone. Hence the tensile force in the luff is

$$\frac{EA_1}{l_1} (l_3 \sin \beta)\theta$$

where EA_1 is another 'effective' axial stiffness which depends on the way the luff is deployed.

Our second equation of motion requires a statement of moment equilibrium. With respect to Figure 5.12, moments about the base of the boom (which is a pin connection) yields

$$\tfrac{1}{3} m l_3^2 \ddot\theta + M l_3^2 \ddot\theta + \left(\frac{EA_1}{l_1} (l_3 \sin \beta)\theta\right) l_3 \sin \beta = \frac{EA_2}{l_2 + l_3} [\Delta_2 - (l_3 \cos \alpha)\theta] l_3 \cos \alpha$$

Hence, in matrix form the equations of motion for free vibration are

$$\begin{bmatrix} I & 0 \\ & \\ 0 & M_2 \end{bmatrix} \begin{Bmatrix} \ddot\theta \\ \\ \ddot\Delta \end{Bmatrix} + \begin{bmatrix} l_3^2\left(\dfrac{EA_1 \sin^2 \beta}{l_1} + \dfrac{EA_2 \cos^2 \alpha}{l_2 + l_3}\right) & -\dfrac{EA_2 l_3 \cos \alpha}{l_2 + l_3} \\ & \\ -\dfrac{EA_2 l_3 \cos \alpha}{l_2 + l_3} & \dfrac{EA_2}{l_2 + l_3} \end{bmatrix} \begin{Bmatrix} \theta \\ \\ \Delta \end{Bmatrix} = \begin{Bmatrix} 0 \\ \\ 0 \end{Bmatrix}$$

185

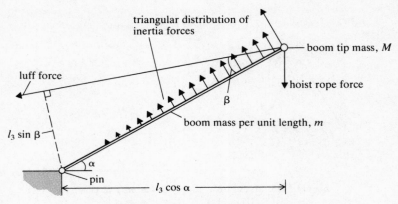

Figure 5.12 Forces acting on boom.

where

$$I = \tfrac{1}{3} m l_3^3 + M l_3^2 = \text{moment of inertia of boom.}$$

These equations can be simplified in a rather convenient way by changing θ to Δ_1, the vertical movement of the tip of the boom. Since $\Delta_1 = (l_3 \cos \alpha)\theta$, the equations of motion become

$$\begin{bmatrix} M_1 & 0 \\ 0 & M_2 \end{bmatrix} \begin{Bmatrix} \ddot{\Delta}_1 \\ \ddot{\Delta}_2 \end{Bmatrix} + \begin{bmatrix} (K_1 + K_2) & -K_2 \\ -K_2 & K_2 \end{bmatrix} \begin{Bmatrix} \Delta_1 \\ \Delta_2 \end{Bmatrix} = \begin{Bmatrix} 0 \\ 0 \end{Bmatrix}$$

where

$$M_1 = (\tfrac{1}{3} m l_3 + M)/\cos^2 \alpha, \qquad M_2 = M_2$$

$$K_1 = \frac{E A_1 \sin^2 \beta}{l_1 \cos^2 \alpha}, \qquad K_2 = \frac{E A_2}{l_2 + l_3}$$

In effect we have reduced the model to the simple case of springs and masses with which we started the work of this chapter.

Now consider the data pertaining to an example drawn from practice. We have

$\alpha \ = 45°$
$\beta \ = 22.5°$
$m \ = 0.4 \text{ t/m}$
$M \ = 5 \text{ t}$
$M_2 = 30 \text{ t}$
$l_1 \ = 40 \text{ m}$
$l_2 \ = 60 \text{ m}$
$l_3 \ = 40 \text{ m}$

The arrangement for the hoist rope system is as in Figure 5.13. The rope is of 40 mm nominal diameter, the effective value of Young's modulus is $1 \times 10^8 \text{ kN m}^{-2}$ based

186

on a net area of 0.62 of the gross area. We need to work out the stiffness of this rope system. Because the hoist is in five falls, a load T gives rise to $T/5$ in the ropes. The stretch is

$$\frac{T}{5}\frac{l_2}{EA} + \frac{1}{5}\left(\frac{T}{5}\frac{l_3}{EA}\right)$$

The stretch is also given by

$$\frac{T(l_2 + l_3)}{EA_2}$$

So

$$\frac{l_2 + l_3}{EA_2} = \frac{l_2}{5EA} + \frac{1}{25}\frac{l_3}{EA}$$

or

$$\frac{EA_2}{l_2 + l_3} = \frac{EA}{l_2/5 + l_3/25}$$

Now

$$EA = 10^8 \times 0.62 \times \frac{\pi}{4} \times (40)^2 \times 10^{-6}$$

$$= 7.79 \times 10^4 \text{ kN}$$

Hence, the equivalent stiffness of the hoist rope system is

$$\frac{EA_2}{l_2 + l_3} = \frac{7.79 \times 10^4}{60/5 + 40/25} = 5.73 \times 10^3 \text{ kN m}^{-1}$$

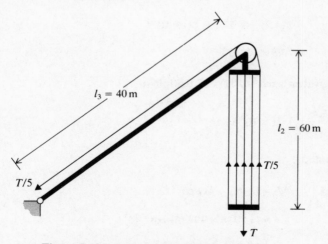

Figure 5.13 Arrangement for the hoist rope.

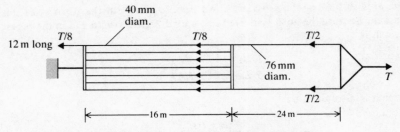

Figure 5.14   Arrangement for the luff.

In the case of the luff the arrangement is as shown in Figure 5.14. The stretch under a load T is

$$\frac{T}{2}\frac{24}{EA'} + \frac{T}{8}\frac{16}{EA''} + \frac{1}{8}\left(\frac{T}{8}\frac{12}{EA''}\right) = T\left(\frac{12}{EA'} + \frac{2}{EA''} + \frac{12}{64\,EA''}\right)$$

Now

$$EA'' = 7.79 \times 10^4 \text{ kN}$$

and

$$EA' = \left(\frac{76}{40}\right)^2 \times 7.79 \times 10^4 \text{ kN} = 2.81 \times 10^5 \text{ kN}$$

Hence

$$\frac{Tl_1}{EA_1} = T\left(\frac{12}{2.81} \times 10^{-5} + \frac{2}{0.779} \times 10^{-5} + \frac{12}{64 \times 0.779} \times 10^{-5}\right)$$

$$= T(4.27 + 2.57 + 0.24) \times 10^{-5}$$

$$= 7.08 \times 10^{-5}T$$

and the equivalent axial stiffness of the luff is

$$\frac{EA_1}{l_1} = 1.41 \times 10^4 \text{ kN m}^{-1}$$

Finally, we find

$$M_1 = (\tfrac{1}{3}ml_3 + M)/\cos^2 \alpha$$

$$= (\tfrac{1}{3} \times 0.4 \times 40 + 5)/\cos^2 45°$$

$$= 20.7 \text{ t} = 20.7 \text{ kN s}^2\text{m}^{-1}$$

188

$$M_2 = 30 \text{ t} = 30 \text{ kN s}^2 \text{m}^{-1}$$

$$K_1 = \frac{EA_1 \sin^2 \beta}{l_1 \cos^2 \alpha} = \frac{1.41 \times 10^4 \times \sin^2 22.5°}{\cos^2 45°}$$

$$= \frac{1.41 \times 10^4 \times 0.146}{0.5}$$

$$= 4.13 \times 10^3 \text{ kN m}^{-1}$$

$$K_2 = \frac{EA_2}{l_2 + l_3} = 5.73 \times 10^3 \text{ kN m}^{-1}$$

STATIC ANALYSIS

For purposes of comparison with peak dynamic quantities we need to know what the static actions are. Figure 5.15 shows the loads. From the matrix equation we have

$$\begin{bmatrix} (K_1 + K_2) & -K_2 \\ -K_2 & K_2 \end{bmatrix} \begin{Bmatrix} \Delta_{1,st} \\ \Delta_{2,st} \end{Bmatrix} = \begin{Bmatrix} 5g + 16g/2 \\ 30g \end{Bmatrix} = \begin{Bmatrix} 128 \\ 294 \end{Bmatrix}$$

or

$$\begin{bmatrix} 9.86 \times 10^3 & -5.73 \times 10^3 \\ -5.73 \times 10^3 & 5.73 \times 10^3 \end{bmatrix} \begin{Bmatrix} \Delta_{1,st} \\ \Delta_{2,st} \end{Bmatrix} = \begin{Bmatrix} 128 \\ 294 \end{Bmatrix}$$

The solution is

$$\Delta_{1,st} = 0.102 \text{ m}, \qquad \Delta_{2,st} = 0.153 \text{ m}$$

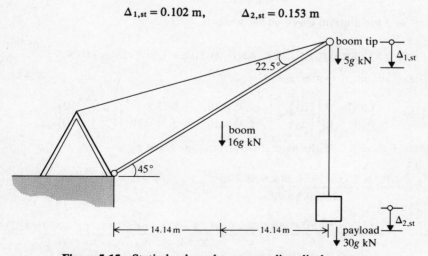

Figure 5.15 Static loads and corresponding displacements.

189

The luff stretches by an amount

$$(l_3 \sin \beta)\theta = \frac{\sin \beta}{\cos \alpha}(l_3 \cos \alpha)\theta$$

$$= \frac{\sin \beta}{\cos \alpha}\Delta_{1,st}$$

and the luff force is

$$T_{1,st} = \frac{EA_1}{l_1}\frac{\sin \beta}{\cos \alpha}\Delta_{1,st}$$

$$= 1.41 \times 10^4 \times \frac{\sin 22.5^\circ}{\cos 45^\circ} \times 0.102$$

$$T_{1,st} = 778 \text{ kN}$$

This force may be found directly from statics, namely

$$T_{1,st}l_3 \sin \beta = 16g \times 14.14 + 5g \times 28.28 + 30g \times 28.28$$

$$= 2219 + 1387 + 8323 \text{ kN m}$$

$$= 1.193 \times 10^4 \text{ kN m}$$

or

$$T_{1,st} = \frac{1.193 \times 10^4}{40 \times \sin 22.5^\circ} = 779 \text{ kN}$$

and we have a useful check on the analysis.

NATURAL FREQUENCIES AND MODES OF VIBRATION

The equations of motion are

$$\begin{bmatrix} 20.7 & 0 \\ 0 & 30 \end{bmatrix}\begin{Bmatrix} \ddot{\Delta}_1 \\ \ddot{\Delta}_2 \end{Bmatrix} + \begin{bmatrix} 9.86 \times 10^3 & -5.73 \times 10^3 \\ -5.73 \times 10^3 & 5.73 \times 10^3 \end{bmatrix}\begin{Bmatrix} \Delta_1 \\ \Delta_2 \end{Bmatrix} = \begin{Bmatrix} 0 \\ 0 \end{Bmatrix}$$

From Equation 5.15 the natural frequencies are given by

$$\omega_{1,2}^2 = \frac{20.7 \times 5.73 \times 10^3 + 30 \times 9.86 \times 10^3}{2 \times 20.7 \times 30}$$

$$\mp \frac{[(20.7 \times 5.7 \times 10^3 + 30 \times 9.86 \times 10^3)^2 - 4 \times 4.13 \times 10^3 \times 5.73 \times 10^3 \times 20.7 \times 30]^{1/2}}{2 \times 20.7 \times 30}$$

190

$$= \frac{4.14 \times 10^5 \mp [(4.14 \times 10^5)^2 - 5.878 \times 10^{10}]^{\frac{1}{2}}}{1.242 \times 10^3}$$

$$= \frac{4.14 \times 10^5 \mp 3.36 \times 10^5}{1.24 \times 10^3}$$

Hence

$$\omega_1 = 7.93 \text{ rad s}^{-1}, \qquad \omega_2 = 24.6 \text{ rad s}^{-1}$$

The natural periods are 0.79 s and 0.26 s, respectively.
The first mode has components

$$a_{21} = 1$$

$$a_{11} = \frac{K_2}{(K_1 + K_2) - \omega_1^2 M_1} = \frac{5.73 \times 10^3}{9.86 \times 10^3 - 20.7 \times 62.9}$$

$$= 0.669$$

Check: From Equation 5.22

$$K_2 a_{11} = (K_2 - \omega_1^2 M_2) a_{21}$$

left-hand side $= 5.73 \times 10^3 \times 0.669 = 3.836 \times 10^3$

right-hand side $= 5.73 \times 10^3 - 62.9 \times 30 = 3.843 \times 10^3$

so the check is satisfactory.
The second mode has components

$$a_{22} = 1$$

$$a_{12} = \frac{K_2}{(K_1 + K_2) - \omega_2^2 M_1} = \frac{5.73 \times 10^3}{9.86 \times 10^3 - 604.8 \times 20.7}$$

$$= -2.154$$

Check: from Equation 5.22

$$K_2 a_{12} = (K_2 - \omega_2^2 M_2) a_{22}$$

left-hand side $= -5.73 \times 10^3 \times 2.154 = -12.34 \times 10^3$

right-hand side $= (5.73 \times 10^3 - 604.8 \times 30) = -12.41 \times 10^3$

so the check is satisfactory.

191

Rather than keep the modes in the above form, it is convenient to normalise them so that

$$[A]^T [M] [A] = [I]$$

or

$$\{a_1\}^T [M] \{a_1\} = 1, \qquad \{a_2\}^T [M] \{a_2\} = 1$$

or

$$c_1^2 \times 20.7 \times 0.669^2 + c_1^2 \times 30 \times 1 = 1$$

$$c_1^2 (20.7 \times 0.669^2 + 30 \times 1) = 1,$$

giving rise to

$$c_1 = 0.160$$

Similarly

$$c_2^2 (20.7 \times 2.154^2 + 30 \times 1) = 1$$

or

$$c_2 = 0.089$$

Hence the normalised modes are

$$\{a_1\} = \begin{Bmatrix} 1.07 \times 10^{-1} \\ 1.60 \times 10^{-1} \end{Bmatrix}, \qquad \{a_2\} = \begin{Bmatrix} -1.92 \times 10^{-1} \\ 0.89 \times 10^{-1} \end{Bmatrix}$$

USE OF RAYLEIGH'S QUOTIENT FOR THE FUNDAMENTAL NATURAL FREQUENCY

Recall that the static displacements under self-weight were

$$\Delta_{1,st} = 0.102 \text{ m}, \qquad \Delta_{2,st} = 0.153 \text{ m}$$

Hence an estimate of the first mode shape is

$$a_{11} = \frac{0.102}{0.153} = 0.67, \qquad a_{12} = 1$$

These are very close indeed to the correct values.
From Rayleigh's quotient

$$\omega_1^2 \leqslant \frac{\{a_1\}^T [K] \{a_1\}}{\{a_1\}^T [M] \{a_1\}}$$

192

$$\leqslant \frac{\{0.67 \quad 1\} \begin{bmatrix} 9.86 \times 10^3 & -5.73 \times 10^3 \\ -5.73 \times 10^3 & 5.73 \times 10^3 \end{bmatrix} \begin{Bmatrix} 0.67 \\ 1 \end{Bmatrix}}{\{0.67 \quad 1\} \begin{bmatrix} 20.7 & 0 \\ 0 & 30 \end{bmatrix} \begin{Bmatrix} 0.67 \\ 1 \end{Bmatrix}}$$

or

$$\omega_1^2 \leqslant \frac{2.48 \times 10^3}{39.3} \leqslant 63.1$$

or

$$\omega_1 \leqslant 7.94 \text{ rad s}^{-1}$$

This, as expected, is very close to the correct answer.

DYNAMIC RESPONSE IN HOISTING MODE

A tending vessel brings the payload alongside the platform. As the payload begins to be lifted the tending vessel drops on a wave with constant velocity V. The peak dynamic responses in the hoist rope and the luff are required. The response calculations are in two parts: first, while the payload remains in contact with the tending vessel the displacement $\Delta_2(t)$ is known and given by

$$\Delta_2(t) = Vt$$

The equations of motion are

$$M_1 \ddot{\Delta}_1 + (K_1 + K_2)\Delta_1 - K_2 \Delta_2 = 0$$

$$M_2 \ddot{\Delta}_2 - K_2 \Delta_1 + K_2 \Delta_2 = F(t)$$

where $F(t)$ is the force in the hoist rope, zero initially, but eventually rising to the value $M_2 g$ at the point at which the payload lifts clear of the tending vessel. Because $\Delta_2(t) = Vt$ during this phase of the response $M_2 \ddot{\Delta}_2 = 0$ and the second equation provides the means by which $F(t)$ is determined, namely

$$F(t) = K_2(Vt - \Delta_1)$$

The first equation becomes

$$M_1 \ddot{\Delta}_1 + (K_1 + K_2)\Delta_1 = K_2 \Delta_2 = K_2 Vt$$

subject to initial conditions of $\Delta_1(0) = \dot{\Delta}_1(0) = 0$.
The general solution is

$$\Delta_1(t) = \frac{K_2}{K_1 + K_2} Vt + A \cos \omega t + B \sin \omega t$$

193

where

$$\omega = [(K_1 + K_2)/M_1]^{\frac{1}{2}}$$

Applying the initial conditions leads to

$$\Delta_1(t) = \frac{K_2}{K_1 + K_2} \left(Vt - \frac{V}{\omega} \sin \omega t \right)$$

This solution holds up to a time t_* such that the payload just lifts clear of the tending vessel. This occurs when

$$F(t_*) = M_2 g$$

and is given by

$$M_2 g = K_2 [Vt_* - \Delta_1(t_*)]$$

$$= K_2 \left[Vt_* - \frac{K_2}{K_1 + K_2} \left(Vt_* - \frac{V}{\omega} \sin \omega t_* \right) \right]$$

or

$$M_2 g = \frac{K_2}{K_1 + K_2} \left(K_1 Vt_* + \frac{K_2 V}{\omega} \sin \omega t_* \right)$$

At this point the second part of the response begins, and we revert to the full equations of motion, namely,

$$\begin{bmatrix} M_1 & 0 \\ 0 & M_2 \end{bmatrix} \begin{Bmatrix} \ddot{\Delta}_1 \\ \ddot{\Delta}_2 \end{Bmatrix} + \begin{bmatrix} (K_1 + K_2) & -K_2 \\ -K_2 & K_2 \end{bmatrix} \begin{Bmatrix} \Delta_1 \\ \Delta_2 \end{Bmatrix} = \begin{Bmatrix} 0 \\ M_2 g \end{Bmatrix}$$

with 'initial' conditions for each of Δ_1, $\dot{\Delta}_1$, Δ_2, $\dot{\Delta}_2$, which are known. The details are as follows.

Solution for $0 \leqslant t \leqslant t_*$

Given $M_1 = 20.7 \text{ kN s}^2 \text{m}^{-1}$, $M_2 = 30 \text{ kN s}^2 \text{m}^{-1}$, $K_1 = 4.13 \times 10^3 \text{ kN m}^{-1}$, $K_2 = 5.73 \times 10^3 \text{ kN m}^{-1}$, and $V = 2 \text{ m s}^{-1}$. Then

$$\omega^2 = (K_1 + K_2)/M_1 = 9.86 \times 10^3/20.7 = 476$$

$$\omega = 21.8 \text{ rad s}^{-1}$$

and

$$\Delta_1(t) = \frac{5.73}{9.86} \left(2t - \frac{2}{21.8} \sin(21.8t) \right)$$

$$= 0.581 [2t - 0.092 \sin(21.8t)]$$

194

$$\dot{\Delta}_1(t) = \frac{K_2 V}{K_1 + K_2}(1 - \cos \omega t) = 1.16[1 - \cos(21.8t)]$$

To find the elapsed time until lift-off we have to solve

$$30 \times 9.81 = 0.581\left(2 \times 4.13 \times 10^3 t_* + \frac{2 \times 5.73 \times 10^3}{21.8}\sin(21.8t_*)\right)$$

or

$$t_* + 0.0636 \sin(21.8t_*) = 0.0613$$

from which

$$t_* = 0.0265 \text{ s}$$

Thus, as the payload lifts clear, the various displacements and velocities are

$$\Delta_1(t_*) = 0.581[2 \times 0.0265 - 0.092 \sin(21.8 \times 0.0265)]$$

$$= 0.0016 \text{ m}$$

$$\dot{\Delta}_1(t_*) = 1.16[1 - \cos(21.8 \times 0.0265)]$$

$$= 0.188 \text{ m s}^{-1}$$

$$\Delta_2(t_*) = 2 \times 0.0265 = 0.053 \text{ m}$$

$$\dot{\Delta}_2(t_*) = 2 \text{ m s}^{-1}$$

Note as a check

$$K_2(\Delta_2 - \Delta_1) = 5.73 \times 10^3(0.053 - 0.0016) = 294.5 \text{ kN}$$

$$M_2 g = 30 \times 9.81 = 294.3 \text{ kN}$$

which is satisfactory.

Solution for $t > t_*$

The next stage in the calculation requires the solution of

$$\begin{bmatrix} 20.7 & 0 \\ 0 & 30 \end{bmatrix}\begin{Bmatrix} \ddot{\Delta}_1 \\ \ddot{\Delta}_2 \end{Bmatrix} + \begin{bmatrix} 9.86 \times 10^3 & -5.73 \times 10^3 \\ -5.73 \times 10^3 & 5.73 \times 10^3 \end{bmatrix}\begin{Bmatrix} \Delta_1 \\ \Delta_2 \end{Bmatrix} = \begin{Bmatrix} 0 \\ 294 \end{Bmatrix}$$

subject to 'initial' conditions of

$$\Delta_1(0) = 0.0016 \text{ m}, \qquad \dot{\Delta}_1(0) = 0.188 \text{ m s}^{-1}$$

$$\Delta_2(0) = 0.053 \text{ m}, \qquad \dot{\Delta}_2(0) = 2 \text{ m s}^{-1}$$

195

The normalised matrix of mode shapes is

$$[A] = \begin{bmatrix} 1.07 & -1.92 \\ 1.60 & 0.89 \end{bmatrix} \times 10^{-1}$$

and

$$[A]^{T} = \begin{bmatrix} 1.07 & 1.60 \\ -1.92 & 0.89 \end{bmatrix} \times 10^{-1}$$

Let

$$\begin{Bmatrix} \Delta_1 \\ \Delta_2 \end{Bmatrix} = [A] \begin{Bmatrix} \xi_1 \\ \xi_2 \end{Bmatrix}$$

where ξ_1 and ξ_2 are the normal co-ordinates of the problem. Since the mode shapes have been normalised by requiring

$$\{a_1\}^{T}[M]\{a_1\} = 1, \qquad \{a_2\}^{T}[M]\{a_2\} = 1$$

we have

$$\ddot{\xi}_1 + \omega_1^2 \xi_1 = \{a_1\}^{T} \begin{Bmatrix} 0 \\ 294 \end{Bmatrix} = 47.0$$

$$\ddot{\xi}_2 + \omega_2^2 \xi_2 = \{a_2\}^{T} \begin{Bmatrix} 0 \\ 294 \end{Bmatrix} = 26.2$$

where

$$\omega_1 = 7.93 \text{ rad s}^{-1}, \qquad \omega_2 = 24.6 \text{ rad s}^{-1}$$

The initial conditions for the normal co-ordinates are found from

$$\{\Delta\} = [A]\{\xi\}$$

or

$$[A]^{T}[M]\{\Delta\} = [A]^{T}[M][A]\{\xi\}$$

or

$$\{\xi\} = [A]^{T}[M]\{\Delta\}$$

since

$$[A]^{T}[M][A] = [I].$$

196

So, in the present case

$$\xi_1 = \{a_1\}^T [M] \begin{Bmatrix} \Delta_1 \\ \Delta_2 \end{Bmatrix} = 2.22\Delta_1 + 4.80\Delta_2$$

$$\xi_2 = \{a_2\}^T [M] \begin{Bmatrix} \Delta_1 \\ \Delta_2 \end{Bmatrix} = -3.97\Delta_1 + 2.67\Delta_2$$

Accordingly,

$$\xi_1(0) = 2.22 \times 0.0016 + 4.80 \times 0.053$$

$$= 0.258$$

$$\dot{\xi}_1(0) = 2.22 \times 0.188 + 4.80 \times 2$$

$$= 10.0$$

$$\xi_2(0) = -3.97 \times 0.0016 + 2.67 \times 0.053$$

$$= 0.135$$

$$\dot{\xi}_2(0) = -3.97 \times 0.188 + 2.67 \times 2$$

$$= 4.59$$

The solutions for the normal co-ordinates are

$$\xi_1(t - t_*) = 0.747 - 0.489 \cos[7.93(t - t_*)] + 1.26 \sin[7.93(t - t_*)]$$

or

$$\xi_2(t - t_*) = 0.0433 + 0.0917 \cos[24.6(t - t_*)] + 0.187 \sin[24.6(t - t_*)]$$

for $t \geq t_*$, where $t_* = 0.0265$ s. Since

$$\begin{Bmatrix} \Delta_1 \\ \Delta_2 \end{Bmatrix} = [A] \begin{Bmatrix} \xi_1 \\ \xi_2 \end{Bmatrix}$$

we have the solutions for the physical displacements as

$$\Delta_1(t - t_*) = 0.107\,\xi_1(t - t_*) - 0.192\,\xi_2(t - t_*)$$

$$\Delta_2(t - t_*) = 0.160\,\xi_1(t - t_*) + 0.089\,\xi_2(t - t_*)$$

The quantities that we are interested in are the dynamic tensions in the luff and hoist rope. For the luff this is

$$T_1(t - t_*) = \frac{EA_1}{l_1} \frac{\sin \beta}{\cos \alpha} \Delta_1(t - t_*)$$

$$= 1.41 \times 10^4 \times \frac{\sin 22.5°}{\cos 45°} \Delta_1(t - t_*)$$

197

or

$$T_1(t - t_*) = 7630 \, \Delta_1(t - t_*)$$

The static tension in the luff (under *all* loads shown in Figure 5.15) is

$$T_{1,st} = 778 \text{ kN}$$

and so the ratio of dynamic to static tensions is

$$\frac{T_1(t - t_*)}{T_{1,st}} = 9.81 \, \Delta_1(t - t_*)$$

The peak value of this is the dynamic amplification factor for the luff (relative to its *full* static tension) and this is found to be

$$DAF_1 = 2.35$$

Similarly, the ratio of dynamic to static tensions in the hoist rope is

$$\frac{T_2(t - t_*)}{T_{2,st}} = \frac{5730}{294} \, [\Delta_2(t - t_*) - \Delta_1(t - t_*)]$$

$$= 19.50 \, [\Delta_2(t - t_*) - \Delta_1(t - t_*)]$$

and the peak value of this is the dynamic amplification factor for the hoist rope, namely

$$DAF_2 = 3.42$$

Plots of the results for the DAFs are shown in Figure 5.16 up to a time of about 0.5 s, which encompasses both maximum values. During this time damping will have had very little effect.

Design of the luff and hoist rope would be based on these peak dynamic tensions after a variety of geometries and payloads had been analysed. Similarly, the interplay of axial load and bending moment in the boom could be investigated using the present model but this needs to be carefully done: the starting point would be the free-body diagram shown in Figure 5.12.

The values of DAF_1 and DAF_2 are different and this is, in large part, due to the differing definitions of the static tension in the luff and the hoist rope. Prior to the application of the payload the hoist rope is slack, but the luff has to carry the boom. This static tension is given by

$$\frac{16g \times 14.14 + 5g \times 28.28}{l_3 \sin \beta} = \frac{2219 + 1387}{15.31} = 236 \text{ kN}$$

Hence the static tension in the luff due to the payload is

$$778 - 236 = 542 \text{ kN}$$

198

Hence the dynamic amplification factor for the luff based on a static tension due to payload alone is

$$2.35 \times \frac{778}{542} = 3.37$$

Clearly, there is now little difference between the values of 3.42 for the hoist rope and 3.37 for the luff. If only the first-mode response had been considered, the tension in the hoist rope would have been

$$DAF_2 = 19.50 [\Delta_2(t - t_*) - \Delta_1(t - t_*)]$$

$$= 19.50 [(0.160 - 0.107)\, \xi_1(t - t_*)]$$

or

$$DAF_2 = 1.03\, \xi_1(t - t_*)$$

where

$$\xi_1(t - t_*) = 0.747 - 0.489 \cos[7.93(t - t_*)] + 1.26 \sin[7.93(t - t_*)]$$

The largest possible value of $\xi_1(t - t_*)$ is

$$\xi_{1,\text{max}} = 0.747 + \sqrt{(1.26^2 + 0.489^2)}$$

$$= 2.10$$

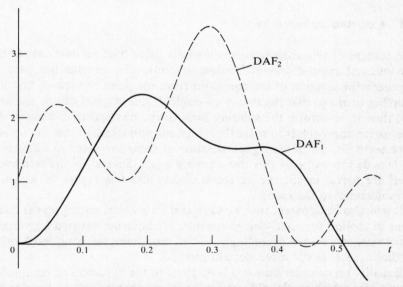

Figure 5.16 Dynamic tensions in luff and hoist rope (DAF_1 and DAF_2 respectively) scaled to corresponding static values.

199

and so

$$DAF_2 = 1.03 \times 2.10 = 2.16$$

Use of the Det Norske Veritas formula yields (see Example 1.1(b))

$$DAF_2 = 1 + \frac{V\omega}{g} = 1 + \frac{2 \times 7.93}{9.81} = 2.62$$

Both these answers are considerably at variance with the result from our 2 DOF model. In the present example, the discontinuities in mass and stiffness are too great successfully to defend use of a single-degree-of-freedom model: the luff is not inextensible and the boom is massive. The second mode contributes significantly to the response.

Finally, and in order to verify the analysis that we have done of the 2 DOF model, note that damping will eventually remove the transient terms in the response and we should be left with the static response alone. In the case of the hoist rope the dynamic amplification factor should be unity when t is large. So, ignoring the sine and cosine terms (they will de damped out with time), we have

$$DAF(t \to \infty) = 19.50 [(0.160 - 0.107) \times 0.748 + (0.089 + 0.192) \times 0.0433]$$

$$= 1.01$$

which is the value expected. This provides a verification of the analysis and, implicitly, supplies unequivocal evidence of the importance of modelling the structure with two degrees of freedom in this case.

5.8 Concluding remarks

The content of this chapter represents little more than an overview of the dynamics of multi-degree-of-freedom systems. The impetus has been to approach the solution of the equations from the point of view of first uncoupling them, so that the theory of single-degree systems can be applied, and then to transform the solution back to the physical co-ordinates. An alternative approach is to solve the equations numerically from the outset. Such methods, which are a generalisation of those presented in Chapter 4, lie outside the scope of this introductory work. Several of the references cited are useful in chasing up these aspects and this course of action is recommended to the reader.

It will also be apparent that we have said little about setting up the equations of motion for multi-degree systems. We have not covered Lagrange's equations, which are exceedingly useful, nor have we considered other methods, such as the finite element method.

Equally, no consideration has been given to the dynamics of continuous systems for which partial differential equations are required to describe the equilibrium of typical elements.

No doubt a case can be made for inclusion of such items, but a line has to be drawn somewhere. In fact, the author's experience has been that, given an understanding of basic structural dynamics, one can construct simple models that adequately reflect structural behaviour. Unnecessary attention to detail is counterproductive.

Another fact is that structural engineers often have to confront problems in structural dynamics after the event. That is, something has gone wrong, or was overlooked in the design stage, and the engineer is then cast in the role of trouble shooter/problem solver. A solution has to be found and found quickly and this forces the engineer to get to the heart of the matter.

The object of structural engineering is to produce structures that are safe, serviceable and economic. Structural analysis and design, of which structural dynamics is occasionally an important part, is simply a means to that end.

Problems

5.1 Find the natural frequencies and mode shapes for the system shown in Figure 5.17.

5.2 Obtain the undamped time histories of response for each mass in the system shown in Figure 5.17 when constant forces, F, are suddenly applied to each mass.

5.3 Rework Example 5.7 when the payload is $M_2 = 40$ t. All other details are to remain the same.

5.4 A portion of a marina pier, which allows small craft to be raised and lowered from the water, is shown schematically in Figure 5.18. An equivalent model is also shown alongside.

Show that the equations of motion for free vibration may be put in the form

$$\begin{bmatrix} (M+m) & m \\ m & m \end{bmatrix} \begin{Bmatrix} \ddot{\Delta} \\ l\ddot{\theta} \end{Bmatrix} + \begin{bmatrix} K & 0 \\ 0 & mg/l \end{bmatrix} \begin{Bmatrix} \Delta \\ l\theta \end{Bmatrix} = \begin{Bmatrix} 0 \\ 0 \end{Bmatrix}$$

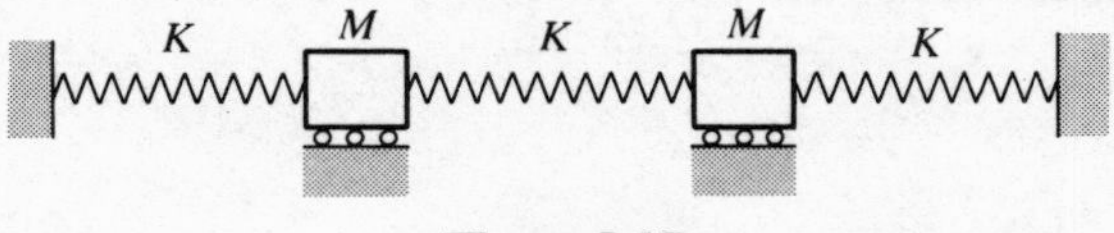

Figure 5.17

201

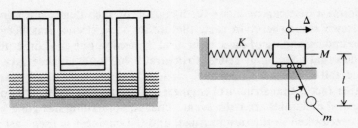

Figure 5.18

5.5 The stiffness matrix (for the three-pin double-arch road bridge shown in Figure 5.19) for the five vertical degrees of freedom is

$$[K] = \begin{bmatrix} 3.33 & 2.31 & -0.75 & 0.73 & 0.73 \\ 2.31 & 3.38 & -1.11 & 0.73 & 0.73 \\ -0.75 & -1.11 & 6.21 & -1.11 & -0.75 \\ 0.73 & 0.73 & -1.11 & 3.38 & 2.31 \\ 0.73 & 0.73 & -0.75 & 2.31 & 3.33 \end{bmatrix} \times 10^4 \text{ kN m}^{-1}$$

The mass matrix is

$$[M] = \begin{bmatrix} 36.7 & 0 & 0 & 0 & 0 \\ 0 & 36.7 & 0 & 0 & 0 \\ 0 & 0 & 36.7 & 0 & 0 \\ 0 & 0 & 0 & 36.7 & 0 \\ 0 & 0 & 0 & 0 & 36.7 \end{bmatrix} \text{kN s}^2\text{m}^{-1}$$

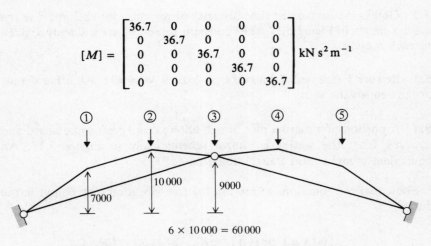

upper chords: $550 \times 350 \times 20$ fabricated plate box section

lower chords: $800 \times 350 \times 25$ fabricated plate box section

Figure 5.19 Double-arch steel road bridge (vertical degrees of freedom shown circled).

202

The flexibility matrix is

$$[K]^{-1} = \begin{bmatrix} 5.80 & -3.88 & -0.08 & -0.27 & -0.25 \\ -3.88 & 5.87 & 0.54 & -0.07 & -0.27 \\ -0.08 & 0.54 & 1.79 & 0.54 & -0.08 \\ -0.27 & -0.07 & 0.54 & 5.87 & -3.88 \\ -0.25 & -0.27 & -0.08 & -3.88 & 5.80 \end{bmatrix} \times 10^{-5}\,\mathrm{m\,kN^{-1}}$$

Use the iterative scheme

$$\{a\}_r = [K]^{-1}[M]\{a\}_{r-1}$$

to show that, after four iterations, $\{a\}_r$ is quite close to the fundamental mode shape of

$$\{a_1\} = \begin{Bmatrix} -6.27 \\ 6.52 \\ 1 \\ 6.52 \\ -6.27 \end{Bmatrix}$$

Hence show that the fundamental natural frequency is,

$$\omega_1 = 16.6\,\mathrm{rad\,s^{-1}}$$

This iterative scheme, which was alluded to in some of the examples in this chapter, can be shown to converge to the fundamental mode. It can be extended to obtain the other mode shapes (and natural frequencies). The technique is called 'the power method' and is widely used. A good reference is the book by Fox (1964, Ch. 9).

References

Arya, S., *et al.* 1979. *Design of structures and foundations for vibrating machines.* Houston: Gulf.

Bathe, K. J. and E. L. Wilson 1976. *Numerical methods in finite element analysis.* Englewood Cliffs, NJ: Prentice-Hall.

Biggs, J. M. 1964. *Introduction to structural dynamics.* New York: McGraw-Hill.

Blevins, R. D. 1981. *Formulas for frequency and modeshape.* New York: Van Nostrand Reinhold.

Brebbia, C. A. and Walker, S. 1979. Dynamic analysis of offshore structures. London: Newnes-Butterworths.

Chopra, A. K. 1981. *Dynamics of structures – a primer.* Berkeley: Earthquake Engineering Research Institute.

Clough, R. W. and J. Penzien 1975. *Dynamics of structures.* New York: McGraw-Hill.

Craig, R. R., Jr 1981. *Structural dynamics: an introduction to computer methods.* New York: Wiley.

Davenport, A. G. 1961. The application of statistical concepts to wind loading of structures. *Proc. ICE,* **20**, 449–72.

Davenport, A. G. 1967. Gust loading factors. *J. Struct. Div. ASCE* **93**. (Reprint No. 457, ASCE Environmental Eng. Conference, Feb 6–9, 1967.)

Dempsey, K. M. and H. M. Irvine 1978. A note on the numerical evaluation of Duhamel's integral. *Earthquake Engng Struct. Dyn.* **6**, 511–15.

Dowrick, D. J. 1977. *Earthquake resistant design.* Chichester: Wiley.

Fox, L. 1964. *An introduction to numerical linear algebra.* Oxford: Oxford University Press.

Housner, G. W. and P. C. Jennings 1982. *Earthquake design criteria.* Berkeley: Earthquake Engineering Research Institute.

Inglis, Sir C. 1963. *Applied mechanics for engineers.* New York: Dover.

Irvine, H. M. 1981. *Cable structures.* Cambridge, Mass.: MIT Press.

Klerer, M. and F. Grossman 1971. *A new table of indefinite integrals.* New York: Dover.

Newman, J. N. 1978. *Marine hydrodynamics.* Cambridge, Mass.: MIT Press.

Newmark, N. M. and W. J. Hall 1982. *Earthquake spectra and design.* Berkeley: Earthquake Engineering Research Institute.

Paz, M. 1980. *Structural dynamics: theory and computation.* New York: Van Nostrand Reinhold.

Peek, R. 1982. *Refined energy balance techniques for the analysis and design of pipe whip restraints.* S.M. Thesis, Department of Civil Engineering, Massachusetts Institute of Technology.

Rayleigh, Lord 1945. *The theory of sound,* vol. 1. New York: Dover.

Simiu, E. and R. H. Scanlan 1977. *Wind effects on structures: an introduction to wind engineering.* New York: Wiley.

Answers

1.1   (a)   570 kN m;

      (b) $$\dfrac{EA_1/(L_1 + L_2)}{1 + (EA_1L_2 \sin \beta \cos^2 \alpha)/[EA_2(L_1 + L_2) \cos^2 \beta]}$$

1.2   0.51 s.
1.5   500 kN m^{-1} per frame, 20 mm.
1.6   0.46 kN m^{-1}, link leg diameter 80 mm.

2.1   320 kN, 0.4 m.
2.2   1.5 m, 0.0054 s, 120 kN.
2.3   16 mm.
2.4   1.2%

3.1   1.08, 1.19.
3.2   105 kN, 41 kN.
3.3   0.67 s, 1.17%, 1.08%.
3.4 $T_h = 2.3$ s, $T_s = 96$ s, $\Delta_h = 0.19$ m, $\Delta_s = 1.8$ m.
3.5   2.7%.
3.6 $M_e = 3400$ kN s^2 m^{-1}, $K_e = 13\,400$ kN m^{-1}, $T = 3.2$ s, DAFs in order: 1.15, 2.09, 5.29, 0.91.

4.1   0.051 m, for a time step of 0.1 s, occurring at $t = 0.5$ s.
4.3   1.04 s, 1.16 s.
4.4   DAF = 1.40.
4.6   0.036 m, 0.67 s.

Index